'Benefitting from being written from both an academic and an insider perspective, this book effectively outlines the challenges that new media pose to the organizational and campaign hierarchy that has traditionally characterized the Conservatives. Ridge-Newman shows that the advent of digital democracy doesn't only pose risks for parties; it also offers rewards.'
– Tim Bale, Professor of Politics, Queen Mary, University of London, UK

'Anthony Ridge-Newman's well-researched book is prime reading for anyone interested in the changing nature of the Conservative Party since the expansion of the internet in British politics. This comprehensive analysis of the role of new media in Conservative organization is balanced, thought provoking and an extremely compelling read.'
–Michael Fabricant, MP

'Anthony Ridge-Newman provides an innovative analysis of the Conservatives' engagement with the internet under David Cameron. Drawing on ethnographic methods during hard-fought elections, his book is unique, breaking new ground in the study of the role that new technologies are playing in the lives of political activists in Britain.'
– Dr Alexander Smith, Assistant Professor of Sociology, University of Warwick, UK

'Ridge-Newman's book offers the first in-depth investigation of the role of digital technologies in the Conservative Party under Cameron. This book explores the challenges that new forms of technology pose to traditional structures of power and authority within political parties. It is essential reading for researchers and practitioners alike.'
– Dr Alex Windscheffel, Lecturer of History, Royal Holloway, University of London, UK

DOI: 10.1057/9781137436511.0001

Other Palgrave Pivot titles

Ian Budge and Sarah Birch: **National Policy in a Global Economy: How Government Can Improve Living Standards and Balance the Books**

Barend Lutz and Pierre du Toit: **Defining Democracy in a Digital Age: Political Support on Social Media**

Assaf Razin and Efraim Sadka: **Migration States and Welfare States: Why is America Different from Europe?**

Conra D. Gist: **Preparing Teachers of Color to Teach: Culturally Responsive Teacher Education in Theory and Practice**

David Baker: **Police, Picket-Lines and Fatalities: Lessons from the Past**

Lassi Heininen (editor): **Security and Sovereignty in the North Atlantic**

Steve Coulter: **New Labour Policy, Industrial Relations and the Trade Unions**

Ayman A. El-Desouky: **The Intellectual and the People in Egyptian Literature and Culture: Amāra and the 2011 Revolution**

William Van Lear: **The Social Effects of Economic Thinking**

Mark E. Schaefer and John G. Poffenbarger: **The Formation of the BRICS and Its Implication for the United States: Emerging Together**

Donatella Padua: **John Maynard Keynes and the Economy of Trust: The Relevance of the Keynesian Social Thought in a Global Society**

Davinia Thornley: **Cinema, Cross-Cultural Collaboration, and Criticism: Filming on an Uneven Field**

Lou Agosta: **A Rumor of Empathy: Rewriting Empathy in the Context of Philosophy**

Tom Watson (editor): **Middle Eastern and African Perspectives on the Development of Public Relations: Other Voices**

Adebusuyi Isaac Adeniran: **Migration and Regional Integration in West Africa: A Borderless ECOWAS**

Craig A. Cunningham: **Systems Theory for Pragmatic Schooling: Toward Principles of Democratic Education**

David H. Gans and Ilya Shapiro: **Religious Liberties for Corporations?: Hobby Lobby, the Affordable Care Act, and the Constitution**

Samuel Larner: **Forensic Authorship Analysis and the World Wide Web**

Karen Rich: **Interviewing Rape Victims: Practice and Policy Issues in an International Context**

Vieten M. Ulrike (editor): **Revisiting Iris Marionyoung on Normalisation, Inclusion and Democracy**

Fuchaka Waswa, Christine Ruth Saru Kilalo, and Dominic Mwambi Mwasaru: **Sustainable Community Development: Dilemma of Options in Kenya**

Giovanni Barone Adesi: **Simulating Security Returns: A Filtered Historical Simulation Approach**

Daniel Briggs and Dorina Dobre: **Culture and Immigration in Context: An Ethnography of Romanian Migrant Workers in London**

DOI: 10.1057/9781137436511.0001

palgrave▸pivot

Cameron's Conservatives and the Internet: Change, Culture and Cyber Toryism

Anthony Ridge-Newman

Department of History, Royal Holloway, University of London, UK

First published 2014 by
PALGRAVE MACMILLAN

Palgrave Macmillan in the UK is an imprint of Macmillan Publishers Limited, registered in England, company number 785998, of Houndmills, Basingstoke, Hampshire RG21 6XS.

Palgrave Macmillan in the US is a division of St Martin's Press LLC, 175 Fifth Avenue, New York, NY 10010.

Palgrave Macmillan is the global academic imprint of the above companies and has companies and representatives throughout the world.

Palgrave® and Macmillan® are registered trademarks in the United States, the United Kingdom, Europe and other countries.

ISBN: 978–1–137–43652–8   EPUB
ISBN: 978–1–137–43651–1   PDF
ISBN: 978–1–137–43650–4   Hardback

This book is printed on paper suitable for recycling and made from fully managed and sustained forest sources. Logging, pulping and manufacturing processes are expected to conform to the environmental regulations of the country of origin.

A catalogue record for this book is available from the British Library.

A catalog record for this book is available from the Library of Congress.

www.palgrave.com/pivot

DOI: 10.1057/9781137436511

In loving memory of my late mother, Regina Marie Ridge-Newman 1957–2007

DOI: 10.1057/9781137436511.0001

Contents

DOI: 10.1057/9781137436511.0001

Preface and Acknowledgements

This book examines the role of the internet in the British Conservative and Unionist Party (also referred to as Cameron's Conservatives (2005–14), the Conservative Party, the Conservatives, the party, the Tories and the Tory Party), with a focus on the party under the leadership of David Cameron 2005–14. In 2005, internet technologies for many people involved interface with primarily Web 1.0, like email and websites. By 2008, Web 2.0, which includes interactive networks of social media, like Facebook and Twitter, had begun making significant technocultural impacts in Britain and, as this book argues, within Cameron's Conservatives. Being a researcher with interests in cultural history and ethnographic methods, observing the advent of these events, as they evolved, fascinated me.

In 2005, I became a registered user of Facebook while studying at the University of North Carolina. At that time, joining Facebook was limited to members of certain universities in the USA. As an early user, Facebook was simply a way for me to keep in touch with my North Carolinian university friends. It was an especially useful communication tool because it was well suited to the peripatetic nature of my life at that time. On moving to Florida, Facebook enabled me to interact daily with groups of friends in North Carolina across a great geographical distance. We could record and share all manner of our lives textually and pictorially, and felt closer and more connected through the use of the medium. Moreover, it was relatively cheap, easy, quick and simple to organize visits in the offline world with large groups of friends in North Carolina while resident in Florida (and vice versa).

On my return to the UK, I was surprised to see the use of the social medium evolving before my eyes. In September 2006, Facebook became open to anyone with an email address. As more and more friends, and groups to which I had links, including the Conservative Party, began using Facebook and other social media in new ways, I began to develop an interest in how these new media were impacting in society. However, when I began the journey that led me to research this book, which developed from earlier doctoral work, I had no way of knowing that I would gain such rich and comprehensive access to Tory intraparty dynamics. Therefore, I would like to give a special thank you to the Conservative Party and its participants for the opportunities presented to me for interaction within the party.

The Runnymede, Spelthorne and Weybridge Conservative Group offered me significant support. I offer thanks to the chairmen of the group for the opportunities they helped me develop, which provided rich interactions in the local Conservative community and led to my role as a councillor for Virginia Water. I am grateful to the central party for giving me the opportunity and honour of joining the Conservative Party's list of approved parliamentary candidates, which led to my being selected as the Conservative Parliamentary Candidate for Ynys Môn | Anglesey in the 2010 General Election. I extend my thanks also to Anglesey Conservatives for selecting me as their candidate and for such a warm welcome to the island. I offer my thanks to my parliamentary mentor, David Jones MP, former Secretary of State for Wales, for his accessibility and guidance.

I would like to thank my friends and colleagues in the party, who have significantly contributed to this book in providing testimonies about their participation in the 2010 General Election. I have chosen to anonymize these contributions for ethical reasons. However, the anonymity should not diminish my gratitude to them for their significant contributions. I would like to acknowledge their willingness and openness to assist the research. Their frank and honest testimonies exceeded my expectations. Many of the interviews were conducted using complimentary rooms at the Carlton Club, St James, for which I am grateful to the club.

I remain grateful to my academic colleagues who have offered me continued support throughout the writing of this book. I give special thanks to my mentor, Dr Alex Windscheffel, for his dedicated advice

DOI: 10.1057/9781137436511.0002

and support. I owe special thanks to Dr Alexander Smith and Professor Tim Bale who have inspired my work. I am especially grateful to them for their ongoing guidance and mentoring. I give thanks to colleagues at Royal Holloway, University of London, who include Susanne Stoddart, Debra Atkin, Dr Francesca Chiarelli, Professor Nathan Widder, Professor Ben O'Loughlin and Professor Andrew Chadwick. I would also like to thank the Economic and Social Research Council and the Friendly Hand Fellowship for funds that enabled the research on which this book is based.

On a more personal note, there are close family and friends to whom I owe my gratitude. Many will go unnamed, so an all-encompassing thank you to them. I must extend my thanks to my very supportive friends Andrew and Davina Palmer; Christopher and Victoria Williams; David and Lorri Newton; the Racusin and Young families of California; and Sandra and Simon Bates. My love and thanks to my wonderful Worcestershire family, especially my uncle, Barry Ridge; my grandparents, Barry and Barbara Ridge; my brother and his family, Ian, Pamela, Harvey and Darcey Newman; and, finally, but most notably, my late mother Regina Marie Ridge-Newman for her 'endless love'.

Bransford, Worcestershire

DOI: 10.1057/9781137436511.0002

List of Abbreviations

Acronyms

AM	Assembly Member (Member of National Assembly for Wales)
BBC	British Broadcasting Corporation
CCHQ	Conservative Campaign Headquarters
CF	Conservative Future
CFers	Conservative Future Activists/Members
CMC	computer-mediated communications
CPC2012	Conservative Party Conference 2012
CUCA	Cambridge University Conservative Association
CWF	Conservative Way Forward
GE2001	General Election 2001
GE2005	General Election 2005
GE2010	General Election 2010
GE2015	General Election 2015
KCL	Kings College London
LBC	London's Biggest Conversation (radio station)
LSE	London School of Economics and Political Science
MEP	Member of European Parliament
MERLIN	Managing Elector Relationships through Local Information Networks (Tory database)
MP(s)	Member(s) of Parliament
MyBO	My Barak Obama (website)
OUCA	Oxford University Conservative Association
PAB	Parliamentary Assessment Board
PDF	Portable Document Format

DOI: 10.1057/9781137436511.0003

PPC	Prospective Parliamentary Candidate
RSWCG	Runnymede, Spelthorne and Weybridge Conservative Group
RSCF	Runnymede and Spelthorne Conservative Future (later RWCF)
RSVP	répondez, s'il vous plait (please respond)
RWCA	Runnymede and Weybridge Conservative Association
RWCF	Runnymede and Weybridge Conservative Future (former RSCF)
TRG	Tory Reform Group
SMS	Short Message Service (text messaging)
TV	Television
UCL	University College London
UK	United Kingdom of Great Britain and Northern Ireland
US(A)	United States (of America)
Y2K	Year 2000 (Millennium Bug)

Contractions and political parties

Conservative Party	
Conservatives	
Tory Party	Conservative and Unionist Party (UK)
Tories	
The party	
BNP	British National Party (UK)
Labour	Labour Party (UK)
Lib Dems	Liberal Democrats (UK)
Plaid Cymru	Party of Wales (UK)
UKIP	United Kingdom Independence Party (UK)

DOI: 10.1057/9781137436511.0003

palgrave▶**pivot**

www.palgrave.com/pivot

1

Cameron's Conservatives and the Internet

Abstract: *This introductory chapter sets the scene for the rest of the book which examines Cameron's Conservatives and the internet. The chapter questions the impact of the internet in the Conservative Party in terms of how it arrived, assimilated and developed in the party's organizational culture 2005–14. It is argued that the use of a more holistic and cultural frame for the analysis of parties and media provides richer elucidations of phenomena. The chapter reviews a range of scholarly perspectives on the Conservative Party. It is noted that between 1997 and 2005, the Conservatives lagged behind their political competitors in terms of e-participation; and how, under the new leadership of David Cameron, since 2005, the party's approach to the internet appears to have advanced significantly.*

Keywords: Cameron's Conservatives; Conservative Party; Cyber Toryism; new media; social media; the internet

Ridge-Newman, Anthony. *Cameron's Conservatives and the Internet: Change, Culture and Cyber Toryism.* Basingstoke: Palgrave Macmillan, 2014. DOI: 10.1057/9781137436511.0004.

The Conservative Party endured 13 years in opposition, 1997–2010, while 'New Labour' presided over a cultural transition to a new millennium. From the illusive 'millennium bug' in the year 2000 (Y2K), to the first release of Apple's 'iPad' in 2010, the New Labour period was partly characterized by developments in computer-mediated communications (CMC). Amid the backdrop of an increasingly complex, transient and globalized world economy, the internet – a dynamic and ever evolving international network of computerized digital communication that has allowed the development of user-led interactive multimedia technologies, for the exchange of commerce, communication, entertainment, information, learning and social interaction – has facilitated the virtual compression of time and space; and greater freedom, choice and access to information for the individual and collective groups (Green 2002; Youngs 2009; van Dijck 2012). In a new millennium characterized by advances in digital technologies, one question for scholars of the contemporary Conservative Party has been: how and to what extent has the party changed under the leadership of David Cameron? (Bale 2008). This book joins a number of wider works in the pursuit of the essence of that question (Snowdon 2010a; Snowdon 2010b; Bale 2010; Dorey et al. 2011; Heppell and Seawright 2012).

In the early 2000s, academic interest in how the internet might have been impacting on party change, notably, at that time, within the Labour and Liberal Democrat parties (Lusoli and Ward 2004), was beginning to emerge. However, there have been few published studies that explore directly and comprehensively the impact of the internet on driving change in the culture of 'Cameron's Conservatives' (Bale 2006). This book attempts to go some way in addressing the gap in the scholarly literature with its examination of the role of the internet in the party's organization, 2005–14. It consists mainly of an analysis of the party's culture in the run-up to General Election 2010 (GE2010), which, for the purpose of this book, begins from the point at which David Cameron became leader of the Conservative Party in December 2005. Lilleker and Jackson's study (2010) of Web 2.0 tools used by six party websites in GE2010 found that the political parties used differing strategic approaches to internet technologies. Therefore, comparisons with other parties are not made extensively in this book, because it would not significantly further enlighten an understanding of how internet technologies impacted on change in the specific case of the Conservative Party.

DOI: 10.1057/9781137436511.0004

This book is largely a descriptive and explanatory study that sits between historical party analysis and the contemporary analysis of parties and new media. The work is influenced by cultural history and ethnographic methodology. It aims to understand culture change using insider knowledge of the Conservative Party. Furthermore, it aims to provide a more focused understanding of the role of the internet in any potential 'decentralization' of Conservative Party processes, rather than party policy. The book's central aim is to build the case for the discovery of a new and somewhat latent technologically-fuelled organizational subculture observed within the Conservative Party between 2008 and 2010. In the specific case of the British Conservative Party, I call this integration of technocultural and political phenomena 'Cyber Toryism'. The nomenclatorial inspiration for this came from Helen Margetts' (2006) evocative model for 'cyber parties'. Margetts presents the cyber party as a new party ideal type, which she argues to fit well the nature of party development in Britain. However, this book does not prescribe to the notion of party ideal types, but rather views them as useful indicators which, when integrated with a party's unique cultural context, provide useful theoretical signposts for the identification of phenomena in relation to a specific party's organization and culture.

Study of media and political parties

The theoretical foundations on which this book rests are influenced by the well-established central thesis of British liberal media history which states that the 'process of democratization was enormously strengthened by the development of modern mass media' (Curran 2002: 4). Historically, advances in the democratic process in Britain, like the five major extensions of the right to vote, between 1832 and 1928, were accompanied by significant developments in mass communications, like the supposed freeing of the press in the eighteenth century; and the growth in film and radio in the early twentieth century. These major developments occurred prior to universal suffrage in 1928. Therefore, this book is based on the assumption that, in terms of media power, the advents of the more recent mass communication technologies, like television and the internet, both of which developed in a period characterized by an historic peak in enfranchisement, have been unrivalled in their potential for impact as tools for democratic and political activity.

DOI: 10.1057/9781137436511.0004

The intraparty dynamics, in other words internal interrelations, of British political parties, like the Conservative Party, can be viewed culturally. Internally, where the different organizational groups and factions interact, divisions and unifications of practices and values are identifiable (Bourdieu 1991). Between these dynamics, symbolic forms of communication are exchanged (Geertz 1973). Traditionally, few political scientists engage with methods that embrace the complex dynamics of political parties in a holistic cultural context (Baynard de Volo and Schatz 2004). However, with prominent work like that of Philip Howard (2006; 2010) and Darren Lilleker (2013), which places the role of new media in the wider context of changing political cultures, political science is recognizing more frequently the importance of culture. This book is influenced by Howard's idea that, firstly, technology can evolve and, secondly, that it has the power to impact on individuals and groups. This supports the assumption, on which the argument for this book is based, that certain technocultural evolutions in wider society can lead to new technological innovations which have the potential to impact at micro- and macro-cultural levels in the Conservative Party. This book aims to explore this theme with a focus on the latent intraparty culture of Cyber Toryism through comparing different groups within the Conservative Party.

Scholars of political parties and political history have tended to divide themselves into subfields that address areas like political communication, party organization, party systems and party development. Political communication tends to focus on the marketing strategies that political parties use to connect with the electorate. Traditionally, scholars of party organization have been interested in the structural components of political parties (Lamprinakou 2008). Party system theory and party development have tended to a focus on generic party models and ideal types (Margetts 2006). Political histories often provide valuable panoramic views of the most salient aspects in the chronology of a party (Ball 1998; Ball and Seldon 2005; and Charmley 2008), but tend to focus on the upper echelons of party dynamics.

The outcome of these sometimes divergent approaches to the study of political phenomena has meant that our understanding of parties can be fragmented. Therefore, some often latent political phenomena have been neglected in academic research and scholarly literature. This book aims to take a more integrated and holistic approach to the study of internet technologies and the Conservatives. A further disciplinary influence that may be evident in this book is the anthropological methodology of New Ethnography. The research on which the book is based was especially

DOI: 10.1057/9781137436511.0004

influenced by the work of Alexander Smith (2011), a socio-cultural anthropologist who conducted an ethnographic study of the Scottish Conservatives. The aim for using an ethnographically inspired approach is to capitalize on my first-hand accounts and experiences generated while embedded in the field with Cameron's Conservatives in the run-up to GE2010 and beyond. While maintaining a critical stance, the aim is to embrace personal and emotional responses in order to unearth what feels 'anthropologically strange' (Hammersley and Atkinson 2009) about contemporary Conservative Party culture.

The interdisciplinary nature of this book means that it draws upon aspects of a range of scholarly disciplines in the social sciences and, as such, does not claim to adhere to any particular discipline in absolute terms. Rather, it seeks to borrow deliberately selected aspects of appropriate theories and concepts, which in some way relate to the study of the role that specific internet-fuelled technologies have played in the Conservative Party's evolving organization and culture. Max Weber's significant works in political sociology, in which he 'was less concerned...to analyse the historical structure of the state than to clarify the nature of the political phenomenon in general' (Thakur 2006: 2) has influenced the thinking behind the approach that this book takes to understanding the Conservative Party. In line with the Weberian view, it is therefore appropriate to identify Conservative Party characteristics, such as its responses to the advent of new media, in order to determine what is significant about the party's nature and evolution. Rather than extensively comparing multiple political parties, the book compares multiple groups and factions within the Conservative Party 2005–14. There is value in a focused case study because it can unearth the interactive dynamics of party phenomena, while allowing the study to be placed within wider time specific contexts (Lawson 1994). Furthermore, being mindful of the Conservative Party's status as a prominent and elite institution in the history of modern Britain, this case study of party change presents a symbolic opportunity to include in the analysis a reflexive sensibility that helps inform our understanding of what such institutional change tells us about wider technocultural change in Britain.

New media and political parties

Before the widespread consumption of television in the 1950s, political strategy focused on the more simplistic forms of advertising and marketing (Wring 2007). Since the early 1970s, emerging technologies, namely

DOI: 10.1057/9781137436511.0004

computers, have been utilized to manage and process political information. At that time, scholars were interested in how parties were employing such technologies, for example the discovery that there had been a shift from handwritten letters to the use of word processing on a significant scale. Recently, we have witnessed new technological phenomena, namely the internet, impact significantly on society and culture (Dahlgren and Gurevitch 2005). However, some scholars warn that researchers should be wary of overstating the internet's significance (Downey and Davidson 2007). This book positions itself in a similar mindset and, therefore, attempts to balance and contextualize its claims through setting the findings within wider historical and societal contexts.

In terms of political culture in Western liberal democracies, the advent of the internet has had its most notable impact in the US, first arriving to widespread prominence with the Obama campaign's approach to online fundraising in the run-up to the 2008 Presidential Election (Vaccari 2010). This and other events have led many to view America as the home of web-campaigning. Furthermore, there has been greater academic interest in such phenomena in the US, with British scholarship remaining within what is currently viewed as a burgeoning field. Consequently, much of the work in the area of new political communications is centred on the United States and, in particular, American elections (Downey and Davidson 2007; Anstead and Chadwick 2009).

In Britain, the earliest mainstream political party websites were launched in the mid-1990s. The Conservatives launched their first website in 1995, a year behind Labour. However, such web presence is considered to have had only a minor impact on the 1997 General Election (Gibson and Ward 1998). At that time, general internet availability in Britain was limited when compared to the subsequent advances in broadband; and mobile and wireless technologies. Therefore, extensive investment in political internet technologies was not deemed as an election priority prior to 2001. In the run-up to the General Election 2001 (GE2001), market segmentation strategies, which were traditionally designed for mailing-out the direct marketing of hardcopies of political communications via the postal service, were beginning to be strategically applied to internet and mobile driven technologies like emails and text messages respectively (Ward and Gibson 2003). However, these developing technologies remained in a supporting capacity in terms of political campaigning.

Based on these trends and those observed in the US 2004 Presidential Election, General Election 2005 (GE2005) was expected to be hailed as

DOI: 10.1057/9781137436511.0004

Britain's first internet election. However, the internet remained secondary to more traditional methods of electioneering (Ward 2005). There was no great qualitative advance in the impact of the internet in elections from GE2001 to GE2005 (Downey and Davidson 2007). Moreover, in between the two elections, the Conservatives lagged behind the other two major British parties in terms of its use of the internet for campaigning. Downey and Davidson's (2007) review of British political party websites considers the main parties' web content and format to have been generic; and that, at that time, the Liberal Democrats were leading the way in the online practice of political blogging.

Between GE2005 and GE2010, the Conservative Party underwent notable change (Bale 2010). Under the new leadership of Cameron, the party attempted to detoxify its dated 'nasty party' image and rebrand itself as an electable and progressive alternative to New Labour. Cameron's contemporary style of leadership, while leader of the opposition, involved the use of internet applications like WebCameron, a video blog. This has been cited as the first significant case of e-politics in Britain (Downey and Davidson 2007). However, Ward et al. have suggested that rather than for use in political marketing, 'internet-based technology might have a greater impact internally within parties' (2005: 27). In part, it is this hypothesis that influenced the development of the research on which this book is based.

By 2005, the internet, as a tool for daily organization, had been assimilated significantly throughout British society and had grown to play a more significant role in the personal and professional lives of ordinary individuals (Livingstone 2005). In the cultural context, 'evolution' and 'technology' are often cited together, especially when researchers write about the rapid advances in 'computerized systems for socializing' (Hofstede et al. 2010: 471). In addition to Margetts (2006), there have been a number of additional scholars who have speculated about how the internet could empower grassroots participation in more networked and less centralized models of political organization (Bimber 1998; Pickerill 2003; Lofgren and Smith 2003). However, this book is not simply interested in analysing general trends in the use of social networks and other web applications for campaigning, rather it is focused on ascertaining whether the Conservative Party's internal culture responded to the technological changes observed in wider culture and, if so, how it manifested itself in its organization and culture, thus providing an elucidation of Cyber Toryism. The book

DOI: 10.1057/9781137436511.0004

analyses whether changes were driven from the top or bottom of the party; and what the implications were and are for the central party and participants at the grassroots.

From 2005 to 2009, access to the internet in the UK rose by 18 per cent, from 55 per cent to 73 per cent (ONS 2006; Ofcom 2010). In the run-up to GE2010, like in 2005, there was significant excitement that 2010 might mark the first internet election in Britain, but instead the internet did not live up to expectations (Gibson et al. 2010a). In terms of general political communication in GE2010, one could be forgiven for overlooking the internet, and questioning whether, in fact, television was the actual new political medium in Britain at that time. Arguably, in some respects, GE2010-style political television was indeed a new medium in British election culture. It was the first general election in British history in which the party leaders went head-to-head in an American-style leader debate (Chadwick 2010). In terms of academic interest in political communications, the television debates have gazumped thus far the historical prominence of the internet in the campaign (Kavanagh and Cowley 2010; Wring and Ward 2010; Bailey 2011; Coleman 2011; Coleman et al. 2011; Lawes and Hawkins 2011).

Literature on the relevance and activity of the constituency campaign in GE2010 is relatively thin. Fisher et al. (2011) found that in general local campaigns used a mix of new and traditional campaign techniques, but the internet was generally low on the agenda. This corresponds with findings outlined in Chapter 7. Rachel Gibson (2010) explores the role of Web 2.0 in the autonomy of participant activity across the main parties in GE2010 and argues that social media has begun to empower activism, but not to the degree observed in the US. Perhaps the reasons for this lie in the different political communication contexts and cultures inherent to the British and American cases, resulting in the US having a more fertile environment than the UK for e-politics (Ward 2005; Gibson et al. 2010b; Williamson et al. 2010). In the Australian case, Gibson and McAllister (2011) found that there were greater electoral benefits to having a candidate website than using social media in the 2010 election. Works such as these tend to address and compare the overarching activities of parties in Britain and other Anglophone political environments. Relatively few works address in detail the role of the internet in the Conservative Party's organizational culture 2005–14.

DOI: 10.1057/9781137436511.0004

Investigating Cyber Toryism

The central problem addressed by this book is to understand what impact the advents of specific internet technologies have had upon the British Conservative Party. The word 'advent' is important in terms of exploring the research questions, because the book seeks to elucidate an understanding of how internet technologies, as new media, have arrived, assimilated and developed in Conservative Party organization and culture, 2005–14. The intraparty dynamics are considered to be organizational phenomena within the internal environment of both the local and national Conservative Party in England and Wales, which this book refers to as the party's organizational culture. Each presented case study is deemed to be unique and individual in terms of its countless variables. The book understands political phenomena to be also socio-cultural phenomena that have no predefined cultural trajectories; but are rather influenced by variables like their own organizational characteristics and external environmental factors.

The term 'impact' is considered to be the repercussions and consequences of phenomenological events; and the role that they play in cultural aspects and the nature of Conservative Party organization. The impact of these events can be limited to an individual (person) or a small collective (group), or be wide reaching for the party, or range on a scale anywhere between the two proportions. The work is built on the basis that certain new mass media, which arrive in a socio-cultural context, arrive in the organizational culture of political parties in different ways in time and space (Tsatsou 2009). It uses both particular/minutia and general/overarching cultural themes. Therefore, the historical impacts of internet technologies in each case study are treated as being organizationally and culturally unique.

The fragmented nature of internet-driven multimedia and multipurpose technologies has come to, in itself, symbolize the general nature of communication technologies in recent times. Therefore, the analysis of the internet technologies in this book addresses each technology on a case basis in order to highlight further the fragmentation that is observable within internet-based channels of communication. That said, some internet technologies are designed to not only interface with people but to also interface with similar technologies. These technologies often share some characteristics and it is, therefore, useful to group some of

DOI: 10.1057/9781137436511.0004

the different types of applications and technologies into categories. For example, early internet applications like email and websites can be categorized as 'Web 1.0'. Facebook, Twitter and blogs can be grouped together as 'social media' or 'Web 2.0'. Social media technologies are characterized by their interactive nature and publically viewable exchange of information in networked multimedia communities online (van Dijck 2013).

In addition to these internet-driven technologies and applications, other internet-based technologies addressed in this book include the Conservative Party's more centralized applications like 'WebCameron', 'MyConservatives' and 'MERLIN' (Managing Elector Relationships through Local Information Networks). These technologies were internet-linked applications that were designed and built for Conservative Party use in party organization and campaign contexts. The book presents and compares the sets of empirical evidence that centred on such technologies in order to develop a deeper understanding of how the Conservative Party has responded to the advent of these notable internet-mediated communications.

The primary hypothesis is that the advent of the internet has impacted on the Conservative Party's historic and elite power structures in that it has loosened aspects of the party's long-established hierarchal organization; and facilitated a degree of cultural empowerment in the technologically savvy cohorts at the party's grassroots. Empowerment is considered to be the active growth of an individual or group to act more autonomously, cognitively, effectively and independently. The concepts of tightening, or party 'centralization', and loosening, or party 'decentralization', are used. Southern and Ward's study (2011) of the impact of the internet in the campaigns of the five main British political parties in GE2010 uses similar concepts. They conclude that new web-based applications, like social media, provided a veneer of localism and, therefore, gave the appearance of a general trend towards decentralization. Moreover, their research found that any general decentralization from internet use was 'countered' by the increased centralization of party databases. This and similar research makes for useful comparison in subsequent chapters.

The aim is to provide the reader visualization tools with which to build pictures that illustrate the strength of grip held by the Conservative Party's leadership and central hierarchy, or elites, over the wider party organization, also referred to as the grassroots. This is particularly in terms of giving indications to shifting power dynamics between elites and the grassroots participants. The term 'power' is understood to be the ability of an individual, or collective, to influence and/or impact on the

DOI: 10.1057/9781137436511.0004

roles of others and/or anthropogenic factors. The hypothesis is used in an attempt to determine the following: (1) whether the advent of the internet, as a potential driver of party change (Bale 2012), impacted on the party's modes of adaptation; (2) whether these were deliberate top-down drivers of change versus organic change driven from the bottom-up; (3) the influence, if any, of such change on the party's evolution; and (4) what such impacts mean more generally for the party's organizational culture, and the distribution of power within the party hierarchy. The use of the term 'hierarchy' is in relation to party organization and the systemization of individuals into an organizational structure that functions in relation to the levels of importance and 'power' (Panebianco 1988) that may be wielded by any given individual or collective.

Providing an absolute and predefined list of variables that constitute the party's culture would defeat the purpose of this study. However, it is always useful in any cultural study to develop an awareness of the types of holistic indicators that might be addressed in the work (Adoni and Mane 1984). Not all indicators were predefined before the commencement of the research, but those which were include Conservative Party: attitudes, behaviours, beliefs, bureaucracies, cultures, customs, innovations, lifestyles, motives, perceptions, resources, structures and symbols. The analysis involves also the assessment of some key historical Conservative Party characteristics in relation to its evolution (Seldon and Ball 1994). These include Conservative Party: activism and engagement (Whiteley et al. 2002); age and demographics (Whiteley et al. 2002); adaptability (Seldon and Ball 1994); association autonomy (Ball 1994a); awareness and consciousness; deference (Ball 1994b); discipline; geographical locations; hierarchies (Seldon and Ball 1994); ideas and ideologies (Heppell and Hill 2005); leadership (Bale 2012); organization; pragmatism (Seldon and Ball 1994); reaction to new technologies (Seldon and Ball 1994); and traditions. The chapters of the book tend to focus around some main fixed objects of interest like Conservative Party: artefacts; affiliated groups; cadres; cohorts; factions; leaders; and participants, all of which are addressed as individual units for analysis.

Ethnographically inspired approach

This book is informed and supported by evidence and data that were collected both on- and off- line while I was in the field with Cameron's

DOI: 10.1057/9781137436511.0004

Conservatives as a committed party participant between 2006 and 2014. Therefore, the research on which this book is based benefits from an insider's perspective and my personal access to an extensive national network of participants within the Conservative Party. The term participant is used as the preferred term throughout the book because it describes both members and supporters of the Conservative Party and recognizes the value that some non-members bring to the party's operations. For example, the party's national network of leaflet deliverers, many of whom will do voluntary work for the party but are not necessarily 'paid-up' members (Fisher et al. 2013).

I have held a number of roles within the Conservative Party, which include being a Conservative Future (CF) branch chairman and president; Conservative councillor; and Conservative parliamentary candidate. These, in addition to other official and unofficial roles within the party, provided me with opportunities to gain access to new and alternative data, materials and sources that do not appear to have been represented before in the academic context. For example, in the run-up to the May 2009 local government and European elections, I began a participant observation in the electorally Conservative county of Surrey, with a focus on the Runnymede, Weybridge and Spelthorne Conservative Group (RWSCG). The RWSCG consisted of the two autonomous Conservative associations whose MPs represented the Runnymede and Weybridge and the Spelthorne constituencies. As will be evident, the ethnographically inspired fieldwork snowballed significantly to present me with many other opportunities for interaction within the party at a variety of levels throughout its hierarchy and in a range of geographical, socioeconomic, political, and cultural contexts. It included a period in the run-up to GE2010 when I was selected as the Conservative Prospective Parliamentary Candidate (PPC) for the Labour-Plaid Cymru marginal constituency of Ynys Môn | Anglesey.

The evidence presented in this book is a result of intensive experiential research from within the field of Conservative politics and elections. Therefore, the work draws on a wide range of sources in addition to the researcher's observations and memoirs. These include: unpublished documents; published documents; articles and communications; formal and in depth semi-structured interviews with Conservative Party participants; and information supplied directly by anonymous Conservative insiders, which are referred to as informants/respondents; or, where and when appropriate, assigned a generic label for general identification of the source like, for example, 'activist' and/or indication of the geographical and organizational relevance of the data, for example 'Anglesey Conservatives'.

In keeping with other ethnographic-based studies (Smith 2011), source details are provided in reference to textual evidence and interview data, but the personal narrative is presented without specific references to the ethnographic source, for example, research logs/journals. Informant and respondent identities are generally protected using anonymity, but some prominent party individuals in well-known public roles are named in order to note their specific roles in pertinent events in the party's recent history.

I informed overtly various Conservative Party participants about the research on which this book is based, including the interviewed respondents with informed consent; and the party board at Conservative Campaign Headquarters (CCHQ). However, as my roles within the party developed, my authentic commitment to the roles that I held within the party meant that it was impractical to inform every individual inside and outside the party with whom I had contact in the course of carrying out my academic, organizational and political roles. Therefore, I have consciously sanitized aspects of the narratives in this book that are informed by anonymous informants in order to ensure that the information disclosed is both ethical and appropriate to the research aims (Segall 2001). In the course of this research, I developed good relationships with a large number of informants. I am committed to maintaining their anonymity.

The manner in which I address this is followed using Alexander Smith's (2011) approach to protecting 'identities'. In most cases, I believe it to be ethical to maintain the informant's anonymity. However, where an individual has acted in an official and public role in the political process, I have engaged my judgement on a case-by-case basis in terms of the extent to which I reveal their actions, sentiments and identities. In 2011, I performed nine detailed and extensive semi-structured interviews of key Tory participants. The positions they have held in relation to the Conservative Party can be found in the bibliography. Each respondent's interaction with the Conservative Party in the run-up to GE2010 is considered to be unique. Collectively, these data inform qualitatively rich indications about what was happening in relation to the internet and the Conservative Party prior to the dates of interview.

DOI: 10.1057/9781137436511.0004

Ethnography, Conservatives and the internet

Philip Howard is a proponent of creatively adapting ethnographic methods for research in e-politics and political culture. 'As new forms of social organization and communities appear, researchers must adapt their methods in order best to capture evidence' (Howard 2006: 208). He describes ethnographic approaches as 'the systematic description of human behaviour and organizational culture based on first hand observation' (Howard 2006: 208). His innovative 'network ethnography' is an integration of network analysis and ethnographic methods. This is an example of how researchers have become increasingly creative in order to tackle the challenge of understanding the cultural implications of the rapid developments in internet technologies. Howard argues for 'a more cultural analytic frame that allows one to treat singular innovations and acts as conditions and symbols of important cultural change in the way we conduct our politics' (Howard 2006: 206) and for 'a more sensible analytical frame' that 'treats technological innovation as co-evolutionary with organizational behaviour' (Howard 2006: 205–6). This idea fits neatly in the context of this book which is based on the assumption that developments in technology can impact on parties like the Conservative Party, thus triggering changes in the evolution of its organizational culture.

Smith's ethnographic study (2011) of a Conservative association in the run-up to the 2003 Scottish and local elections is focused on describing and understanding the culture of its bureaucracy and activism, and how the Scottish Conservatives interacted in social and political contexts. In this sense, it has some similarities to the approach taken by this research and is, therefore, a useful reference point. However, conversely, this book places the social and political contexts in the background and puts the view of the party's relationship with new media in the foreground. Smith states that his academic interests in the party began before he became actively involved within it. In contrast, I became a Conservative member at least two years before considering and conducting this study. Therefore, as my political interests predate my academic interests, Smith and I have approached the ethnographic study of the Conservative Party from opposing ends of the same plane.

Smith's study pays some attention to the traditional communication practices at the Conservative association level, such as the use and symbolism of the widely used Conservative medium of the 'InTouch' leaflet. Although there is further mention of the use, coordination, and quality, of CMC technologies in the campaign, the analysis of the role that new media played is not a focus of his work. The setting is also quite different. Since Scottish devolution, Conservatism north of the border has underperformed when compared with the Conservative Party's progress in England and the devolved electoral region of Wales (Smith 2011). Therefore, the wider political backdrop in which this book is set is different in temporal and geographic terms because it presents primary data that is concerned mainly with the Conservative Party in England and Wales 2005–14.

Ethnography can illuminate what other approaches do not consider relevant. There are significant parts of everyday cultures that go unnoticed by those living in them, and the positivistic methods that are often employed to research them. Conversely, organizational ethnographers seek to draw out the intricate everyday aspects of the organizational environment (Koot 1995). Approaches to studying political histories, organizations and cultures, tend to, like Howard's work, address the more salient issues. Therefore, much of the everyday mechanics which collectively power the political machines of our democracy remain latently unrecorded. Conversely, this book seeks to connect the everyday practices of the ordinary party participant to the more elite and prominent, some might say glamorous, aspects of Conservative Party culture and life.

Overview of this book

As discussed earlier in this chapter, the post-Thatcher Conservative Party changed most notably from 2005 under the new leadership of David Cameron. An outward example of such change was the launch of the ground breaking WebCameron video blog in 2006. Chapter 2 examines the impact of WebCameron and two other Tory internet-linked technologies, MyConservatives and MERLIN. The chapter focuses on the role of Conservative Party elites and CCHQ operations in the development of these new Tory technologies. It is suggested that when the party leader demonstrates change, wider party change follows (Charmley 1996; Taylor 2008). On this basis, the chapter argues that the innovative WebCameron platform was a symbolic event that catalysed further internet use in the wider party and led the way in terms of greater participation in e-politics in the party culture and beyond. While supposedly giving greater interactive access to the real politician, WebCameron

DOI: 10.1057/9781137436511.0004

was rather more the embodiment of top-down change. Therefore, the advent of WebCameron is deemed to be a demonstration of the party seizing more absolute control of its own communication output through a particular communication technology, the likes of which had not been seen since the early days of TV in the 1950s.

In the run-up to GE2010, according to the Total Politics Political Blog Directory, the Tories dominated the political blogosphere in terms of the number of blogs associated with the Conservative Party. Research by Southern and Ward (2011) supports this observation. Chapter 3 explores some prominent themes and pertinent examples of phenomena in the Tory blogosphere, including the potential pitfalls of using the microblog platform, Twitter. The chapter focuses on the place of the ConservativeHome blog in contemporary Conservative organization. The chapter particularly examines the role of prominent individuals like ConservativeHome's founder, Tim Montgomerie. It is identified that Montgomerie and others benefitted from raised profiles through engagement in Cyber Toryism and subsequently filled niches in cyberspace that elevated them to a new elite status. These phenomena are compared with the roles of central elites in Chapter 2. It is argued that ConservativeHome is an example of Cyber Tory leadership that helped catalyse change from the grassroots upward and, in turn, it steered aspects of central party change in the form of greater internal transparency.

Chapter 4 provides an alternative narrative of Cyber Tory activity in that it shifts the spotlight from more elite figures in the party to those at the grassroots who are perhaps lesser known in the public sphere. The chapter is rooted in an analysis of the role of Facebook in the party's organizational culture from 2008 onward. It is argued that while the phenomena in chapters 2 and 3 were significant in influencing a proliferation of uptake of the uses of new media at the party's grassroots, the culture of Facebook participation evolved naturally through a learning and copying behaviour. The chapter reveals the importance of leadership in this process and identifies specific individuals in the ranks of the younger cohorts of the party as significant influencers of change. The chapter argues that Cyber Tory Facebook participation was characterized by a technologically centred innovation culture that helped dissolve traditional geographical and hierarchical barriers to grassroots activity.

In 2014, it might seem a more normalized concept to imagine signing up to an organization using the internet. In fact, for many it would now be preferable in order to reduce bureaucracy, paperwork and postage/

DOI: 10.1057/9781137436511.0004

travel costs. However, in 2006, for many in the Conservative Party this was a relatively new concept. Chapter 5 takes a look at the journey of becoming a Conservative Party member from the participant's perspective. The chapter identifies how the party's online processes were out of sync with its traditional membership structure and that, while the party was in transition in the run-up to GE2010, the party seemed to lose some active engagement potential from its online membership because the party was ill-equipped to convert weaker forms of online membership (Margetts 2006) in to stronger forms of face-to-face participation. The chapter provides the narrative of the ethnographer journeying from an online political neophyte to a fully initiated and active member of Cameron's Conservatives.

Chapter 6 is the first of two geographic case studies that place the use of internet technologies in the Conservative Party within the constituency organization and campaign contexts. The chapter details the first-hand observation of the participation of the researcher and others within the RWSCG context in Surrey. The chapter features an analysis of the role of internet technologies in a local council by-election campaign in 2009 from the candidate's perspective. It is argued that there was both an age and digital divide observable amid cohorts in the Surrey Conservatives and that trust and rapport building were central to dissolving cyber-based barriers to deeper engagement within local Conservative associations. It is found that the internet facilitated new network interactions that made association and campaign organization a looser and more fluid experience, which ultimately led to richer and more diverse campaign-based and social interactions in the offline world.

The second of the geographic case studies features the run-up to GE2010 through the eyes of the researcher as the PPC for Ynys Môn. Chapter 7 is divided into subsections that generally focus on particular groups of communication technologies, like, for example, Web 1.0 and Web 2.0. It is argued that although the internet was used to enhance the campaign in some circumstances, for instance, the use of the medium to facilitate a remote candidate presence, and a virtual campaign team of participants separated by large geographical distances, the internet, in particular e-campaigning, was not a priority for use within the campaign. In fact, although the internet helped with organization and process, the lack of sufficient internet capacity in some aspects of the party's wider national campaign organization led to negative bureaucratic impacts at the local campaign level for the candidate.

DOI: 10.1057/9781137436511.0004

The book culminates in Chapter 8, which is used to present and bring together some of the more recurring and prominent themes from earlier chapters. The chapter is used to develop the case for Cyber Toryism as a cultural singularity in the history of the Conservative Party in existence mainly between 2008 and 2010. It is argued that, before WebCameron, a widespread culture of political-focused internet use in the party context was not evident; and that, post 2010, the use of internet technologies in the party organization became a more normalized practice across cultural and age divides which meant it ceased being a subculture and a more mainstream constituent aspect of wider Tory organizational culture. The chapter concludes that generally the internet acted like a lubricant oiling Conservative Party processes, which in turn resulted in greater fluidity within networks and organizational and campaign operations. It is argued that this loosening of centralized control allowed for shifting power dynamics and subsequent culture change to occur. It is found that the younger cohorts were central to this change in party culture, which had remained more traditional since John Major's leadership; and that the advent of David Cameron's leadership acted to punctuate, in other words speed-up, technocultural evolutions in the life of the wider party.

DOI: 10.1057/9781137436511.0004

2
Tory Elites and Centralized Internet Operations

Abstract: *Chapter 2 examines the impact of WebCameron and two other Tory internet-linked technologies, MyConservatives and MERLIN. The chapter focuses on the role of Conservative Party elites and CCHQ operations in the development of these new Tory technologies. It is argued that the innovative WebCameron platform was a symbolic event that catalysed further internet use in the wider party and led the way in terms of greater participation in e-politics in the party culture and beyond. While supposedly giving greater interactive access to the real politician, WebCameron was rather more the embodiment of top-down change. Therefore, the advent of WebCameron is deemed to be a demonstration of the party seizing more absolute control of its own communication output through a particular communication technology.*

Keywords: CCHQ; David Cameron; MERLIN; MyConservatives; political elites; WebCameron

Ridge-Newman, Anthony. *Cameron's Conservatives and the Internet: Change, Culture and Cyber Toryism.* Basingstoke: Palgrave Macmillan, 2014. DOI: 10.1057/9781137436511.0005.

In the run-up to GE2010, CCHQ, at Millbank, was restructured to incorporate a growing digital media team, which included individuals like Craig Elder, Rishi Saha, Rohan Silva and Samuel Coates, former deputy editor of ConservativeHome. These were in addition to senior communications strategists, who included Steve Hilton and Andy Coulson, former editor of the *News of the World*. As senior aides to the party leader David Cameron, both Hilton and Coulson had significant influence over the Conservative Party message (Bale 2010); and were members of 'Cameron Central', an elite inner circle within Cameron's leadership team (Seawright 2013).

Matthew Hindman (2009) argues that, while some claim the internet to be a democratizing force in politics, digital democracy is a myth and that new elite participants have filled niches in its place. This and other chapters of the book provide evidence that support both sides of the argument, which suggests that the role of the internet in politics is more complex than some researchers might argue. Furthermore, Jon Lawrence (2009) is sceptical about whether internet technologies encourage greater public interaction in political culture and asks questions about how politicians can appear credible when doing so on the internet. Lawrence pertinently asks – 'how do we know it's them?' (2009: 251) and questions the authenticity behind political platforms like WebCameron. That said, in a pre-existing culture in which politicians use communication assistants to write their speeches, is it important for politicians to be the ones hitting send on a blog post? Or is it enough for them to lend their name and face to output that is generated in an ideas culture of backroom elites? This chapter uses first-hand accounts in order to explore more deeply applications like WebCameron and the party elite's approach to the internet. It focuses on the role of WebCameron in the daily life of the party leader, and CCHQ staff, and examines also the two other main centrally controlled Conservative internet technologies – MyConservatives and MERLIN.

WebCameron: watching the Tory leader

Cameron, within a year of becoming the Tory and opposition leader, launched his online campaign in the form of WebCameron at www. webcameron.org.uk on 30 September 2006 (Woodward 2006). The event marked the beginning of a new type of campaigning activity for the leadership of the Conservative Party. Moreover, it was the first use of a video blog by a prominent political leader (Downey and Davidson 2007: 95). As Anstead and Chadwick (2008) note, this event was followed by the use of Web 2.0 by all of the Labour Party deputy leadership candidates in the run-up to the contest, 24 June 2007. This suggests that WebCameron may have led the way in promoting the acceptable use of Web 2.0 among political elites in Britain.

Tim Bale (2010) suggests that Cameron's first WebCameron video, which showed the Conservative leader washing-up dishes, and being a dad at home with the kids, was symbolic in that it encapsulated 'the new man' ideal of the time. The strategic use of WebCameron conveyed a number of additional messages to the electorate. It portrayed the Tory leader, who was well known for his privileged Etonian background (Hill 2013), as a man in touch with the busy work/life balance of contemporary Britain. It suggests that Cameron, as a son of the British elite, realized the importance of conveying an image of normalcy in a time when television viewing was dominated by a growing appetite for 'reality' (Arthurs 2010).

British television and other media in the 2000s had become dominated by a public fascination for observing the mundane nature of everyday lives (Hill et al. 2007). WebCameron, was the Conservative Party's attempt at a web-based, Do-It-Yourself, Tory leadership version of *Big Brother* (see Arthurs 2010: 182). The advent of new internet video technologies, and the growth in new media uses, had enabled the Conservative Party elite to effectively broadcast their leader via his own personal online channel for the first time. Parties and politicians have traditionally worked on a repetitive strategy basis in order to maintain control over the message within their participatory environment (Lilleker 2013). The use of WebCameron gave Cameron and his team complete editorial control over output, the likes of which had not been seen since the early days of political television in the 1950s (Seymour-Ure 1996; 2003). Through WebCameron this could be integrated with the party message in a more visually accessible manner than ever before.

Furthermore, WebCameron was symbolic also in party leadership terms, because, like those prominent new media using Conservative leaders before him, for example Stanley Baldwin's use of film (Taylor 2002) and Harold Macmillan's mastery of television (Bale 2012), Cameron was attempting to master the internet – the new political medium of his time. The use of the internet in this way also came at a time when there had been an acknowledgment of the media's focus on personalities in politics (Norris

DOI: 10.1057/9781137436511.0005

2000). David Seawright's (2013) analysis of 'Cameron 2010' suggests that Cameron himself was well aware of the significance of personality politics and that WebCameron was one of a number of strategies used by the Tories to which Cameron was keen to lend his personality.

According to the first-hand account given by Respondent Two, Cameron and Cameron Central played a significant role in the party's use of new media applications like WebCameron. The respondent describes how Francis Maude, as party chairman, enthusiastically oversaw the party's use of internet technologies in campaigns; and that, within the Tory leadership, George Osborne had the best understanding of, and talent for, the technological uses of internet applications. Moreover, the respondent claims that Osborne 'was obsessed with getting the iPad, which at that point had not been released in the UK'; and took a hands-on approach to driving forward the central party's use of the internet.

This testimony should be tempered contextually with the acknowledgement that the source's role was rooted in promoting the use of digital technologies inside the party. However, at the very least, it indicates that there was an aspirational interest in the possible uses of developing internet-linked technologies within an elite cohort at the top of the party. This apparent interest in technological developments and the intimate working dynamic between Cameron, Maude and Osborne, demonstrates that there was strategic unity within Cameron Central in terms of its appreciation for the potential power and uses of the internet. This is in contrast to their predecessors when compared to the Conservatives' lacklustre approach to the internet prior to 2005 (Chapter 1). Therefore, it is plausible to suggest that this new leadership attitude towards the internet was a significant turning point for the party. In keeping with Hindman (2009), Cameron Central were acting to take control and secure the party's place as the dominant British political elite presence on the internet.

In a party like the Conservative Party, in which, historically, the party leadership has determined the course, hue and identity of British Conservatism in any given period (Taylor 2008), it is important that the party leadership embodies and symbolizes the desired directional course in order for the wider party to adapt in line with any aspirational change (Charmley 1996). The advent of WebCameron symbolically demonstrates Cameron Central's commitment to modernizing the face of the party through the use of new technologies. This is a significant development because of the potential financial costs that generally accompany any investment in the developmental use of new technologies. As suggested

DOI: 10.1057/9781137436511.0005

by the following testimony, support from Cameron Central was necessary in order to secure the initial investment for the successful execution of WebCameron.

> David Cameron and particularly George Osborne were instrumental in empowering our team and giving us the backing that we needed at the top level. Another key person to mention in the 2006 period was Francis Maude. He was absolutely instrumental as party chairman. So, in terms of organizational structure, Francis was absolutely key particularly with the launch of WebCameron, which in 2006 was the first of its kind – as a political party leader's video blog. It did a very good job in terms of humanising David, because David was obviously a massive, massive asset. He was certainly unlike what had come before him in that he was able to connect with the British people in a way that a Conservative leader had not since the early days of Major. So, for us, we needed to be able to use that asset, and David was very keen, and in particular Steve Hilton, who was Cameron's Director of Strategy, was very keen that we used the internet as much as possible (Respondent Two).

There is the internal recognition here that the party had struggled to adapt itself under the leaderships of William Hague (1997–01), Ian Duncan Smith (2001–03) and Michael Howard (2003–05). This is something that has not escaped academic analysis. One of the central questions has been why did the Conservatives 1997–2005 uncharacteristically take so long to brush themselves off and do what they needed to do in order to win elections (Bale 2010)? It would seem that with Cameron's selection as leader, and his subsequent willingness to embrace the new technologies and contemporary issues of the time (like, for example, environmental sustainability and, later, same-sex marriage), the party has reengaged with some form of Tory pragmatism, which, when compared with Conservatism before it, would appear to have been reinvented for this new and globalized millennium (Heppell 2008). That said, the above testimony suggests that the party's decision to change, and its implementation of pragmatic steps to demonstrate such change to the electorate, was centred on the deliberate inclinations and actions of a cohort of a few 'key' elite individuals. This further highlights an elite and highly centralized approach to enacting change within the party, which intensified with the advent of Hague's 1998 'Fresh Future' reforms to the party organization (Bale 2012). Contextually, it is important to note that this has been accompanied by a widespread trend of declining party membership across a range of political parties since the 1960s (Whiteley 2011).

DOI: 10.1057/9781137436511.0005

Whether or not the central party understood it at the time, since the 2010 General Election there has been a certain recognition and awareness within Cameron Central that the act of accentuating Cameron's youthful attributes, and modern man appeal, through WebCameron and the internet, was a vote winner and a significant component of him connecting with a new generation of voters.

> Because the internet is quite useful in a branding sense – we needed to be portrayed to be a younger fresher party that was more in touch with the kind of needs and values in modern Britain. Heavy use of new technology was seen as a way to do that. But more importantly it was seen that we could reach a key demographic, a younger demographic – an internet demographic if you like. There was also a real focus on the idea that the mainstream media was forcing us always to unhelpfully aim for that five-second clip on the *Six O'clock News*. New media allowed us to make our statement, not at length, because apparently nobody wants to listen to a politician talk at length, but it would allow us to have our say on our terms (Respondent Two).

This perspective, from an individual within the team behind the digitizing of Cameron's Conservatives, shows that the use of the internet by the party, in 2006, was a strategic choice stemming from the very top level, with the explicit intention of reaching out to a younger generation of Conservative voters and supporters. Furthermore, the idea that WebCameron would be a tool for communicating the Cameron brand on Cameron Central's 'own terms' was a view also held by a 2010 grassroots party activist. This individual testifies that a number of Tory campaign videos 'went viral' and that WebCameron was 'an invaluable resource to show people what David Cameron thought', from which he could also select 'the moments that he wanted to share with the rest of the world' (Respondent Six).

Through innovation and an embracement of the user-led nature of the internet, political parties in 2006 had the potential to harness and somewhat control aspects of the internet. Political leaders holding the power to exerting significant control over the mass media is something that had been lost since the early days of political television (Kandiah 1995; Lawrence 2009). Notably, the political power of the internet has been recognized by lower ranking politicians from within the Conservative Party. Douglas Carswell (a former Tory MP who recently defected to UKIP) and Daniel Hannan (a Conservative MEP) suggest that the internet can be used to empower also the outside politician, over party bigwigs (Carswell and Hannan 2008). This is not least exemplified

DOI: 10.1057/9781137436511.0005

by Hannan's speech, which castigated Gordon Brown in the European Parliament and went viral on YouTube within hours (Hannan 2010).

Therefore, Cameron Central made an early, some might say astute, choice to attempt to control aspects of the national party brand through the use of the internet. However, there are indeed pitfalls to the use of the internet, especially in terms of publicity in this personality-focused political culture. WebCameron's relatively successful execution is further pronounced when considered against the comparatively unsuccessful attempt by, the then, Prime Minister, Gordon Brown to connect to the electorate using YouTube (Theakston 2011). Therefore, it seems plausible to suggest that success in the use of internet media is based on the appropriateness of the fit, in terms of the natural style and/or the adaptability of any given prominent politician, to the specific medium being used. The internet is not unique in this sense, because, in 2010 television terms, Cameron was considered to have underperformed in the first leader debate when compared to the Liberal Democrat leader Nick Clegg (Lawes and Hawkins 2011). Furthermore, it seems personality was beginning to play a role like in the late 1950s when Macmillan's strategic adaptation of style for the new medium of television began its course (Cockett 1994; Bale 2012).

In 2006, the Conservative Party's use of the internet, as a new medium, in the form of WebCameron, was a work in progress. According to Respondent Two, the understanding of how best to use the medium was learnt through an evolutionary process, which impacted on the manner in which the Conservatives who were associated with WebCameron organized their weekly duties.

> You could not have WebCameron without Cameron. He was always really, really committed to it. The early days were interesting because it was a very amateur production, but that actually gave it a lot of its charm. Four years on, after we made many strides forward in terms of video production quality, we hired a professional camerawoman to come in and work very closely with David [Cameron] – with him, in many cases, for a lot of the time during the day – and people still remember the video that was shot on the tiny handycam, dimly lit in his kitchen. David was always very accessible. He was always very interested in it. He was always challenging us to make it better and make it more interesting. I had many jobs when I first arrived at the party, so David did not always appreciate quite how busy I was, along with the rest of the team. But, quite often, I would be in the back of the car with David going to various places and just chatting away with him and he would ask why I could not be with him all the time (Respondent Two).

DOI: 10.1057/9781137436511.0005

This demonstrates how the party's adaption to the internet was beginning to change the nature of some of the traditional roles of CCHQ staff and, to some extent, place higher demands on their workload. The incorporation of new internet media in the roles of communications staff was impacting in a way that it demanded more versatility; and a conflict between growing on-the-road duties and the need to edit and upload WebCameron content in a technologically suitable environment and manner. Essentially, as campaign and new media trends developed over time, Respondent Two assumed the role of being an in house citizen journalist and cameraperson in addition to performing other, more traditional, roles. Therefore, the inherent challenges embodied in the evolving nature of new media would have made writing a prescriptive political communications job description, or indeed a technologically current party campaign strategy, way ahead of time, somewhat futile. It is appropriate to place these changes in Conservative Party culture in the context of the wider impacts of social media technologies on societal change. Theories of globalization, like that of Gillian Youngs, suggest that there has been a 'blurring of public and private spheres' in 'social, political, economic and cultural life' (2009: 127). The above testimony suggests that Conservative Party culture was not impervious to such external environmental influences in the run-up to 2010.

However, in this case, these contributory factors should be balanced against the role of the personal inclinations of Cameron Central. Respondent Two believes that without the enthusiasm of the key political leaders at the top of the party, the party's organizational approach towards the use of new internet media would have been slower to develop:

> Cameron took a really keen interest in WebCameron which enabled us to do a lot of things. David gave us two things. He looked at the videos and said, "Well why would anybody want to watch this?" He gave us an unprecedented level of access and trust and he also gave us the backing we needed essentially to go off and get CCHQ budget to hire someone with a broadcast background to follow him with a professional quality camera and produce a broadcast quality film – which took WebCameron up to the next level. The reason why that was incredibly important for us was that, come the election, and really in the year leading up to the election, Sky and the BBC would call us and say "We have just seen the WebCameron video; can we have the tape of that, we want to play that out on TV". I do not need to tell you the quality [significance] of having footage controlled by us, put together by us, and broadcast by us on the *Six O'clock News*. That was pretty incredible for us.

DOI: 10.1057/9781137436511.0005

Cameron's approach to the use of new media in his campaign would suggest that when the party leadership takes ownership of a new medium, the medium has a significant chance of gaining some prominence amid competing traditional media platforms. This is, again, reminiscent of the late 1950s and the party's approach to television under Macmillan (Cockett 1994). In the case of WebCameron, it would seem that Cameron himself understood the public mood of the time: that voters wanted access to the real politician, rather than the gloss and 'spin' that had become characteristic of the New Labour years (Esser et al. 2000; Kuhn 2007). It is, therefore, curious that WebCameron should be discontinued and its content removed from easy publically accessible sources like YouTube (Hern 2013; Payne 2013) in the run-up to the 2015 General Election (GE2015). This apparent retraction from the party's approach to using the internet for greater party openness and political transparency has led to questions over the central party's authenticity and commitment to its modernization agenda, since Steve Hilton's departure in 2012 (Payne 2013). It is also a demonstration of the party exerting centralized control over the internet applications that are within its ability to control.

In the run-up to GE2010, the advent of wider social media use in Britain provided Cameron with the opportunity to take some control over his campaign's 'public face' (Bale 2012) and, therefore, portray himself as a grounded politician in-touch with the real lives of real people through reality footage. This demonstrates that Cameron held some personal confidence in the medium; and had a relaxed attitude towards the recording and broadcasting of the more mundane aspects of his life as a political leader. Again, this analysis is in keeping with Youngs' (2009) theory above. Cameron's confident integration of aspects of his private and public life is likely to have been rooted in an appreciation for the control that the Conservatives held over the production of the footage and the manner in which it was broadcast via the internet.

However, although the controlled nature of WebCameron was seen as a positive victory for the Conservative political and organizational elites in the media battle of GE2010, it was not long before the Cameron machine had to adapt again its ideas of how to use effectively the internet as a tool for campaigning. As one Conservative MP testifies, the approach to brand-Cameron loosened overtime as the party began experimenting with the inclusion of more interactive forms of new media, such as having live interactive chats on Mumsnet, a medium over which Cameron had little control (Respondent Nine). Bale (2010)

DOI: 10.1057/9781137436511.0005

suggests that Cameron's Mumsnet webchat was one of a string of public-ity stunts which the party stage managed in order to present a cuddlier-Conservative-face to the public. That may be the case, but it shows that Cameron's Conservatives were prepared to take innovative risks in their use of new media. Furthermore, it demonstrates that the Tories were observing changing trends in new media uses and incorporated some plasticity into their digital strategies in order for their plans to adapt and change over short time scales in line with the rapid developments in new internet applications.

Prior to 2006, the use of the internet by party members had been largely limited to personal and private purposes. However, the party leader's act of using WebCameron, signalled to the wider Conservative Party that it was okay to integrate the use of new internet applications with the activity of communicating the Conservative message. Respondent Two believes that WebCameron was the beginning of a change in the political use of internet technologies in Britain and that it 'kick started more inno-vative use of web-technology'. This suggests that the party leader's use of WebCameron acted as a catalyst for further participation in e-political communication and organization within the party. Respondent Two also claims that WebCameron was the finest hour of the Conservative new media campaign, because it made a positive story on the front page of *The Guardian* during the Conservative Party Conference 2006.

> ...this new era of engagement where politicians were appearing at the hand of a tiny handycam on tiny little YouTube videos. I think that WebCameron probably did play its role in changing the way politics is done online. I would not be so grandiose to suggest that it played as much of a role as anything that happened in the Obama campaign, but I would probably say that it was not far off it (Respondent Two).

The respondent criticizes the WebCameron project for being costlier to the party than necessary, believing that the project could have achieved the same results 'using a free blog platform on a YouTube channel'. Moreover, with the benefit of hindsight, the respondent questions why the party did not simply broadcast Cameron using social media; but suggests that WebCameron, as a platform, allowed the party to do more than it could have achieved using YouTube because the party was able to encourage interaction through the 'Ask David' application. Ask David facilitated online voting to allow the public a choice over which questions were posed to Cameron for his response, and thus placed some control

DOI: 10.1057/9781137436511.0005

of the agenda into the hands of the electorate. Respondent Two estimates that WebCameron received an average of 150,000 hits per day during the early stages of the project; and describes how these numbers eventually settled to between 5000 and 6000 hits per day. This was significantly fewer hits than Conservatives.com – the party's central website. Eventually, in the run-up to GE2010, WebCameron was migrated to Conservatives.com where it became integrated with the party's corporate web presence. This helped to drive web-traffic to both platforms through search engine optimization techniques; and build a diversified audience for the Cameron-Conservative brand (Respondent Two).

WebCameron had its genesis in the mind of Hilton (Respondent Two) and this narrative tells us something about the relationships between Cameron Central and the CCHQ digital team. It demonstrates the elite ideas culture at the top of the party from which the trajectory of the Tory use of new technologies stemmed. The initial idea was fertilized by the enthusiasm of the party's most influential leaders who chose to move the party forward in their embracement of new internet technologies. The WebCameron project evolved over time and the work of CCHQ staff began to adapt and change in order to accommodate new media demands. Ultimately, the WebCameron project facilitated a direct channel for Cameron, as leader of the opposition, to not simply connect, but, also, interact with the electorate in a manner that had not been seen before. Most significantly for the Tories, WebCameron acted as both a symbol and a visual signal, validating the use of the internet in Tory politics. Subsequently, in the years which followed, the wider party's use of internet technologies increased significantly.

By 2010, internet-based democratic activity known as 'citizen-initiated campaigning' had emerged as a trend in British political culture (Gibson 2013). In keeping with this trend, it would appear that the activities of the WebCameron team, for example, the citizen journalist style roles that were undertaken, placed these party actors in roles of being also first-hand actors in the media output process. This is somewhat reminiscent of the way central office staff in the late 1950s adapted their roles to incorporate the use of broadcast technologies in a new TV studio at Conservative Central Office (Cockett 1994; Bale 2012). Therefore, it is evident that, as in the case of television in the 1950s, the impact of internet-based technologies had begun blurring some of the roles, activities and responsibilities of individuals within CCHQ in the run-up to GE2010.

DOI: 10.1057/9781137436511.0005

MyConservatives and MERLIN – campaign magic?

In addition to WebCameron, the two other most significant developments in CCHQ's aspiration for the use of new internet-based technologies in the run-up to GE2010 were (1) MyConservatives, a Web 2.0 application; and (2) MERLIN, an internet-linked database. WebCameron was a Cameron-centric application that was aimed at facilitating interaction between the party leader and general public specifically. Whereas MyConservatives and MERLIN were aimed at integrating both local and central campaigns and supporting party organization through the networking capabilities of the internet.

MyConservatives

MyConservatives was first developed in the run-up to GE2010 and was an interactive online venue aimed at connecting individuals involved in Tory campaigns. The Conservative Party estimates that MyConservatives brought together 10,000 people on 390 campaigns (Conservatives 2012a). The website got off to a troubled start because it crashed immediately after launch. The party claimed that the initial crash was due to an unexpectedly high demand (Chivers 2009). This unsuccessful launch may have contributed to the application's general demise that led to it being deemed a failure (Lilleker 2013). The functionality of the site was modelled on Barack Obama's Web 2.0 campaign website 'MyBO', which was used in his 2008 US Presidential Election campaign (Lilleker and Jackson 2010). MyConservatives was aimed at raising the profile of candidates and promoting an awareness of the issues central to their campaigns. The application was used as a fundraising tool aimed at directly connecting candidates to potentially likeminded electors, who were not necessarily Conservative activists or party members. The use of the internet in this party branded and real time participatory manner was not unique to the Conservatives in run-up to GE2010. According to a study by Lilleker (2013), the British National Party (BNP) website www.bnp.org.uk led the way in terms of user interactivity and participation. Its functionality resembled closer to that of Obama's website in that every page could be commented on and shared via social media. The same study places the Liberal Democrat website www.libdemact.org.uk in a similar league to www.bnp.org.uk in terms of its interactivity and impact.

DOI: 10.1057/9781137436511.0005

In 2012, the party's central website Conservatives.com stated that the party had learned from the use of MyConservatives in 2010 and that the application would be revamped for the run-up to GE2015 and constitute an ongoing tool in the party's online campaign. However, in the run-up to the European Elections, 22 May 2014, the MyConservatives URL on Conservatives.com directed the user to a holding page designed to capture personal data, which included five questions in an electronic survey that would help ascertain an individual's contact details, location and socioeconomic class (Conservatives 2014). This data harvesting strategy, which can be combined with targeted email campaigns, is consistent with growing trends observed in the party's approach to local and national e-campaigns over the last decade (Downey and Davidson 2007).

Craig Elder (2010) claims that the Conservative Party raised £500,000 through online donations. Respondent Two claims that the Conservatives raised approximately 25 per cent of their online fundraising via MyConservatives and that the central purpose of MyConservatives was to encourage online donations from individuals who would be willing to support a specific candidate but not necessarily the Conservative Party directly. The respondent admits that the application would have been more successful if it had been launched a year earlier in September 2008. Respondent Four, a prominent Conservative blogger, agrees on this point and suggests that MyConservatives did not have enough time to mature. This notion fits with Mergel and Bretschneider (2013) theory, which suggest that there is a three stage maturation process in the evolution of organizational institutions that adapt to new modes of communication in the form of established Web 2.0 participation. Furthermore, as Kay McNutt (2014) rightly argues, in many cases it is not always the technology that obstructs the efficacy of Web 2.0 participation, but rather the bureaucracy, culture and organization of the institution.

Respondent Nine, a safe seat Tory candidate in 2010, made some use of MyConservatives, but raised the relatively insignificant amount of £450 for the constituency campaign. The respondent testifies that some of her friends used it to conveniently donate online. As a candidate in GE2010, I noted that Eric Pickles, as party chairman, contacted candidates and encouraged them to use MyConservatives creatively with the aim that large numbers of supporters would donate £1. It would seem that this was an attempt to emulate the fundraising success of the 2008 Obama campaign. In comparison, MyConservatives was a flop. Respondent Nine claims that a small number of people used MyConservatives

DOI: 10.1057/9781137436511.0005

to donate amounts generally ranging between £10 and £50, with one person donating £150; and criticizes the application as being frustrating, 'because none of it really worked'. The respondent testifies that Twitter was used most extensively in their local campaign and, comparatively, MyConservatives failed to make a significant impact.

As a target seat candidate, Respondent Eight testifies that their campaign used MyConservatives in a similar manner to Respondent Nine raising approximately £200. Respondent Eight claims that MyConservatives was used to organize some elements of their campaign and that, although it was an improvement on what the party had before, it was not as streamlined in its functionality or as 'effective' as some online charity fundraising applications like 'JustGiving'. Both respondents agree that MyConservatives was a work in progress for the party and that, with development, the application could have the potential to significantly assist campaign fundraising and organization in the future. Therefore, this suggests that, in the run-up to GE2010, MyConservatives had only a minor impact on the party's organizational culture.

Like WebCameron, MyConservatives attempted to mediate direct and controllable access between politician and audience. However, when compared to WebCameron, the MyConservatives brand was significantly inferior and launched too late gain any significant traction. Recent findings by Pich et al. (2014) suggest that the Conservatives, and indeed other parties, would do well to strengthen their intraparty communications when attempting to develop an external political brand, because it is the party participants who mostly export the brand at constituency level. The respondents' testimonies provide a general consensus that MyConservatives, as a new medium, was a viable concept with potential for the future, but that it simply failed to mature in the eyes of its users in time for it to reach its full potential in 2010.

MERLIN

MERLIN is a centralized database system that was networked nationally in the run-up to GE2010 and beyond. CCHQ had a direct link via the internet to each Conservative association using the MERLIN technology and used this to gain visible access to elector data (Fisher et al. 2011). The associations were required to purchase specific technology in the form of a MERLIN-specific desktop computer in order to use the MERLIN network in conjunction with constituency canvass data. The central party stipulated that the MERLIN computer had to be located in

DOI: 10.1057/9781137436511.0005

a generally accessible association office in order that no one individual could have overall control of the data. MERLIN was the successor to the former Blue Chip database (Chapter 6), which was an autonomous and un-networked computerized system. The main purpose of both systems is to manage canvass and membership information.

According to Respondent Three, chairman of the Runnymede and Weybridge Conservative Association (RWCA) in 2010, MERLIN contains information for 40–50 million electors. The database was another new Tory-specific technology which was widely perceived within the party organization to have failed to reach its potential prior to 2010 (Respondent Two). This was a topic of frustration across many types of individuals involved in the Conservative Party in the run-up to the election (Respondents Two; Three; Eight). Respondent Three suggests that the central party's ability access to the MERLIN database hindered the local association's use of email in the election.

> Because all of the email addresses that are held on MERLIN are available to the party centrally, what we have is the party involved in a communications programme in which the associations are not involved. What I believe they're doing is damaging our ability to use this communication.

Furthermore, according to Respondent Two, MERLIN's inadequacies hindered also the potential uses of MyConservatives:

> MyConservatives ended up being a standalone platform with data which needed to be manually inputted and extracted. That should never have been the case. In 2010, you should have been able to make any database speak to any database and we should have been able to make that work much more effectively than we were ultimately able to do. We certainly lay the blame for that at the feet of the infrastructure problems that MERLIN faced. Could it be that it was a project too big? Could it be that we bit off more than we could chew? For somebody else to answer, but it seems that way to me.

These testimonies seem to support findings by Southern and Ward (2011) that party databases acted as a countering force to wider trends in electronic forms of decentralization (Chapter 1). However, it would also appear that the impact of the Tories' centralized database operations were such that MERLIN was not as effective at taking control of local data as the central party had intended.

Respondent Eight claimed that his campaign team used effectively MERLIN as a successor to Blue Chip, but suggested that future improvements to MERLIN were needed in order to integrate successfully its

DOI: 10.1057/9781137436511.0005

uses with other new internet technologies used by the party, for example tablet and smart phone technologies.

> MERLIN is a much friendlier interface, it is much easier to use and it can provide you with much more relevant and targeted information than the Blue Chip system. The Blue Chip system was getting on for twenty years old and it was beginning to show. I think with MERLIN it was a step forward, but it is still a pretty clunky piece of software and you could see the real difference between the business world in which you have a huge market, and therefore the software gets developed very quickly, and very effectively, and the political world in which the market is actually a lot smaller. Therefore the software is rather slower, more out of date and clunkier (Respondent Eight).

In comparison with Blue Chip, MERLIN was viewed by these respondents as an advance in technology for the Conservative Party. However, their testimonies suggest that as a technology with internet capabilities, MERLIN did not in general change or revolutionize significantly the culture in which Conservative canvassers participated in campaigns. That said, according to Fisher et al. (2011) the speed at which the party could exchange information between local and central MERLIN databases meant that the party could adapt its strategies in key seats. Therefore, although MERLIN did not significantly change internal activities to the extent that WebCameron had, it did act to compress some temporal and geographic factors (Green 2002) in terms of outcomes from data interface between the local and central party.

Procedural customs around the uses of Blue Chip had developed already over a 20-year period prior to the advent of MERLIN. Therefore, in terms of its role in canvassing, MERLIN inherited its user culture from Blue Chip. The internet capabilities of MERLIN in 2010 enhanced the central party's access to database information. However, MERLIN's interface with other Conservative internet-based technologies like MyConservatives was limited due to the infancy of both technologies and the lack of lead time for them to mature prior to the run-up to GE2010. Had there been a speedier development and integration of these technologies then it is likely that their capabilities would have played a significantly greater role in the nature of the party's campaigns and the organizational culture in which party participants used the technologies (McNutt 2014). Unlike, WebCameron, MERLIN and MyConservatives had not been advanced to a stage in which they held any real revolutionary capacity to change the manner in which the party organized itself at the grassroots. However, there is evidence to suggest that some database administrators, like association agents and chairmen, underwent a process of bureaucratic adaptation in order to successfully interface with the user complexities and harness the potential of association databases in innovative ways (Chapter 6).

DOI: 10.1057/9781137436511.0005

3
Blogs: The Conservative Home?

Abstract: *Chapter 3 explores some pertinent examples of phenomena in the Tory blogosphere, including the potential pitfalls of using the microblog platform Twitter. The chapter focuses on the place of the ConservativeHome blog in contemporary Conservative organization. The role of prominent individuals like ConservativeHome's founder, Tim Montgomerie, is examined. The chapter identifies that Montgomerie and others benefitted from raised profiles through engagement in Cyber Toryism and subsequently filled niches in cyberspace that elevated them to a new elite status. It is argued that ConservativeHome is an example of Cyber Tory leadership that helped catalyse change from the grassroots upward and, in turn, it steered aspects of central party change in the form of greater internal transparency.*

Keywords: Conservative blogosphere; Conservative grassroots; ConservativeHome; Cyber Tory; Tim Montgomerie; Tory dissent

Ridge-Newman, Anthony. *Cameron's Conservatives and the Internet: Change, Culture and Cyber Toryism.* Basingstoke: Palgrave Macmillan, 2014. DOI: 10.1057/9781137436511.0006.

There is a significant body of work that documents the trend, since the 1960s, in the decline of political membership in Britain and elsewhere (Whiteley 2011). This change in the mass-based party has given rise to claims that grassroots activism has been eroded (Whiteley and Seyd 1998). However, by 2005, the power of the internet was beginning to be harnessed significantly in political activism and meant that there was a developing realization about the potential for activist empowerment (Lusoli and Ward 2004). Shirky (2008) suggests that audience members are being converted into active participants through more interactive and social media trends. In terms of the Conservative Party's grassroots, new innovations in the use of internet technologies appear to have facilitated new online approaches to political engagement that have been held firmly before the gaze of the central party and its elites.

Gibson et al. (2013) found that blogging can empower party participants through providing 'public voice'. Tim Bale suggests that 'the website ConservativeHome provided an institutionalized forum for complaints – and one that could be easily accessed by the media' (2010: 291). However, as this chapter will discuss, ConservativeHome has become more of a platform than a forum. It has been used to shine a spotlight on certain issues and individuals, thus raising profiles in the wider media. The advent of this grassroots-led Conservative blog has acted to peel back the curtains of the Conservative Party and provide a transparent window through which the public can gaze upon the party's views and mechanisms like never before. This chapter explores the role of the blogosphere in Cyber Toryism and analyses the role of blogs like ConservativeHome. Reference to the 'Conservative' or 'Tory' blogosphere is loose in that it denotes blogs that have a range of connections to the Conservative Party. These include blogs of party participants who may or may not support the status quo of Conservative Party policy and organization, but do have a focused and discernable interest in associating the blog and its contents with rudimentary aspects of Conservative Party identity, which in itself remains a broad coalition of views (Heppell 2013).

Tory blogosphere

The internet being a powerful, sometimes subversive, force in the Conservative Party in the run-up to GE2010 was perhaps most encapsulated in the Conservative blogosphere. Many Conservative blogs could

DOI: 10.1057/9781137436511.0006

be found listed in the *Total Politics* 'Political Blog Directory', which was a comprehensive public listing of blogs within the political blogosphere totalling 2351 blogs on 11 June 2010. The publication linked bloggers to its website and allowed bloggers to self-submit their blog to its listings. According to the list, in June 2010 there were approximately 417 Conservative affiliated blogs, compared with 245 Labour affiliated blogs and 261 Liberal Democrat affiliated blogs. These figures support my observation that in the run-up to GE2010, the Conservative blogosphere was the most abundant and active in terms of the number of active Conservative blogs and the frequency at which the bloggers published articles. Moreover, results from research by Southern and Ward (2011) also support this observation. Their figures for GE2010 show that 58 per cent of Conservative parliamentary candidates used blogs compared to significantly lower figures for Labour at 40.5 per cent and the Liberal Democrats at 32.3 per cent. However, in terms of impact in public and political spheres, in general, the most prominent political bloggers tended to be non-parliamentarians.

At a Conservative Party forum event, in early 2010, the Chairman of the Welsh Conservatives introduced Iain Dale as a speaker and suggested that very few active Conservative participants would go to bed at night without having read the 'Iain Dale's Diary' blog. Whether or not that statement is true, it demonstrates the Conservative Party's appreciation for the role of blogging and that there was recognition in senior party ranks that individual Conservative bloggers could achieve prominence and influence in the daily lives of Conservative participants. In this case, party participants had adapted to the new medium of blogging through integrating their daily reading of traditional media like newspapers in digitally accessing the writings of growing political celebrities like Iain Dale, thus raising the blogger's profile and placing them in a new position of significance both inside and outside the boundaries of traditional party membership. The case also demonstrates the wide reaching impact of some political bloggers, whose blogs were being read by participants from the party's grassroots to its organizational elites.

In January 2009, Harry Cole's 'Tory Bear' blog broke a story involving some Conservative Future activists or 'CFers', also known formally as the Young Conservatives, who had seemingly posted indelicate comments on Facebook in relation to attending a bad taste costume party (Cole 2009; BBC News 2009; Respondent One). Cole's blog post subsequently led to the story being reported on the BBC1 *Six O'clock News*. The resultant

DOI: 10.1057/9781137436511.0006

outcome of the affair culminated in the bad taste CFers being publically suspended by the party. This pitfall in social media engagement was not unique to politics or the Conservative Party. In the run-up to GE2010, a Labour parliamentary candidate published, what was deemed to be in the media, inappropriate comments on Twitter, which, due to a 140 character limit per post, is considered to be a microblog platform. Like the Conservative example above, the Labour Party was quick to silence and remove their party participant from the public gaze (Forsyth 2010).

These examples demonstrate how interacting with new media in the run-up to GE2010 could seal the end of a political career in the time it takes to write and post 140 characters. Since 2009, there have been a string of other similar examples that have involved representatives from parties like UKIP and the BNP. Even in mid-2014, with the use of social media in politics maturing, a Conservative candidate resigned when 'anti-Islamic' and 'homophobic' comments were uncovered on his Twitter feed (Simpson 2014). This suggests that in the run-up to GE2015 there remains within the e-politics community a relative naivety about the pitfalls of not self-sanitizing, what some might deem, politically incorrect views.

The incomparable aspect of the Tory Bear case is the manner in which the BBC catapulted a story written by a, then, little known, Conservative affiliated, blogger into the nation's gaze on its primetime news programme. This was arguably a significant David and Goliath moment in British media history, because the preeminent media giant ran a story that was a scoop broken by a relatively unknown citizen journalist. In the run-up to GE2010, the internet facilitated the potential for the giant to follow the little man and create a pseudo celebrity status within political microcultures. Like Dale's rise to become an LBC Radio presenter, Cole's raised profile presented opportunities for him to leave the Conservative fold. Cole became a co-collaborator in the right wing blog order-order.com, joining forces with the prominent anarchistic blogger Paul Staines (known as Guido Fawkes). The use of blogs in this manner is a significant example of internet technologies facilitating the potential for dissent within political culture (Ward et al. 2005).

In Chapter 2, the case of WebCameron was presented as symbolizing how 'personality-based' politics (Norris 2000; Seawright 2013) seemed to be infiltrating British political culture through the use of internet technologies as early as 2006. In 2014, one case in particular exemplifies how (micro-) blogging can be used to impact on the profile of political elites with, arguably, both positive and negative outcomes simultaneously. Michael Fabricant MP, former Vice-Chairman of the Conservative Party and former Government Whip, has been widely reported to have been given the sack as vice-chairman of the party because of his 'eccentric' comments on Twitter (Fryer 2014; Urwin 2014; Wright 2014). As noted above, this type of outcome has become more or less common place in wider political culture since the run-up to GE2010.

However, what is most interesting about the Fabricant case is his persistent integration of controversial views, personal eccentricities and the apparent use of social media like Twitter in order to relentlessly pole vault himself further into the gaze of the public. In doing so, commentators have reported Fabricant's political 'demise' (Wright 2014) while, at the same time, somewhat affectionately, celebrating his celebrity-like status (Urwin 2014). Fabricant's blunders and misjudgements (Freyer 2014) combined with a twist of personality may have reduced his popularity in elite Tory circles, but the power of dissenting blogging culture to attract an audience can be demonstrated by the former whip's 17,600 Twitter followers (@Mike_Fabricant 2014). Fabricant reports that his weekly reach has exceeded, on occasion, 2.4 million reads and mentions, which he ascertains using the sumall.com web application (Fabricant 2014). Fabricant's online impact is further highlighted when compared to his more traditional elite Tory colleagues like David Jones MP, former Secretary of State for Wales, who has 7,952 Twitter followers (@DavidJonesMP 2014); and Justine Greening MP, Secretary of State for International Development, who has 15,200 followers (@JustineGreening 2014). This case demonstrates the significance of the role of the individual in the outcomes associated with online political activity.

ConservativeHome

In the 2010 narrative, the ConservativeHome blog was another blog that played a significant role in the Conservative Party in the run-up to the election. It was, and remains, the most illustrious example of Cyber Toryism in terms of its role and impact in the party's day-to-day organizational culture and the blog's influential prominence, which reached the attentions of Conservative participants at virtually all levels of the party's hierarchy (Respondents One, Two, Six, Eight and Nine). Respondent Two testifies that the CCHQ Press Department dealt with bloggers in the run-up to GE2010, but suggests that it is important not to 'confuse blogging with journalism'. The same respondent also argues that bloggers, especially those involved with ConservativeHome, behave rather more like lobbyists and pressure groups, even though they may come from a journalistic background. However, another respondent, a former editor of ConservativeHome disagrees:

> I am a journalist. I worked for the BBC for four years. I worked for the *Daily Telegraph* for five years. I have worked for ConservativeHome for two and a half years. As far as I am concerned, I am still pursuing a journalistic career by doing ConservativeHome. I suppose a lot of bloggers would regard themselves as kind of individual citizen journalists (Respondent Five).

The vast majority of bloggers do not have this journalistic pedigree; however, it would seem that the most prominent bloggers in the 2010 Cyber Tory community, like Iain Dale, Jonathan Isaby and Tim Montgomerie, did have significant publishing and journalistic experience prior to becoming blogging idols.

Respondent Five suggests that blogging is a fluid phenomenon which is constantly evolving over time. Therefore, the newness of the role of blogging as in interdisciplinary phenomenon in the run-up to GE2010 meant that a clear consensus was and remains absent among the Conservative sophisticate in terms of a definition of what blogging is and what bloggers are; and how the blogosphere fits in the wider picture amid the traditional institutions of state like the press, the public, pressure groups and political parties. Dan Burstein's (2005) historical anthropology of blogging argues that blogging has been in constant evolution because its precursors dateback to cave paintings and more recently relate to the practice of writing a diary. He argues that these are cultural communication artefacts; and phenomena similar to blogs have been observed repeatedly throughout human history. This suggests that communicating in forms of public record is intrinsic to human culture; and with the advent of the internet, the practice of blogging has moved forward a very natural human characteristic that has been displayed in new forms.

Blogs are significant in the evolution of social media because they bring together an integration of the personas, views and egos of individuals with a diverse range of internet-based technologies and applications (Burstein 2005). This goes some way in explaining why a hierarchy formed in the Conservative blogosphere in the run-up to GE2010 – and why the prominence of specific bloggers have remained established so

DOI: 10.1057/9781137436511.0006

far in the subsequent run-up to GE2015. Hindman's claim (2009) that new elites fill niches in cyberspace would seem to apply in this case. Perspectives in media ecology help explain this further in terms of media hybridization in which new media appear to be integrating technologies, human characteristics and functional applications (Scolari 2012). The most prominent Cyber Tory bloggers already had a significant public profile rooted in a journalistic or publishing background. Therefore, they had a significant base on which to build further their cyber profiles and integrate their personas with new media applications. The use of internet technologies in this way is similar to that taken by the central party's approach to WebCameron, which integrated Cameron's personal and political personas and views in the online video blog project. As these blogging personalities and their resources grow, for example income and staff, questions over credibility should be asked (Lawrence 2009). Because, as Jon Lawrence might ask, how are we to be sure the face of the blog is the individual actually writing its output?

The uniqueness of blogging is its versatility as an internet platform for use in policy and discourse. It can be moulded to be what individuals or groups of collaborators want it to be in order to serve their own purposes. In terms of political culture, it places a personal and individualistic stamp on output – rather than mirroring the identity of a party's tradi-tional elite. Therefore, it is plausible to suggest that it was for this reason that blogging meant different things to different actors in the party and the wider democratic environment. Even with the benefit of hindsight following GE2010, a degree of ambiguity in relation to blogging, and microblogging, for example Twitter, remained extant in Conservative discourse as recently as October 2012.

A *Daily Telegraph* columnist and ConservativeHome critic, speak-ing at the 'Has Social Media Changed the Conservative Party' forum at Conservative Party Conference 2012 (CPC2012), suggests that ConservativeHome is a remarkable phenomenon that has created a new force in politics and is not yet fully understood. The columnist claims that the blog is used to propagate the views of Lord Ashcroft, the blog's funder, and Tim Montgomerie, the blog's founder. The columnist argues this to be a narrow representation of the Conservative sophisticate that attracts 'sharp-suited' lobbyists who represent only '0.001 per cent' of the population. This is in contrast to the observations that I recorded in the run-up to 2010 which suggest that ConservativeHome rapidly and widely spread as a recognizable and trusted brand among a diverse cross

DOI: 10.1057/9781137436511.0006

section of Conservative party participants who regularly referenced the blog in their verbal, written and digital discourse. The columnist also suggest that the online publication is politically positioned to the Right of Cameron's Conservatives' ideology and policy, and thus likened the blog to the 1980s Leftist movements within the Labour Party. Of course, the difference being that political movements in the 1980s had to manage their message through traditional media channels, like for example the *Militant* newspaper, and had no globally accessible, dynamic, low-cost digital technologies from which to broadcast their views. Generally, resource intensive protests that required man power and organization had to be rallied in order to achieve impact. In the case of ConservativeHome in the run-up to GE2010, small groups of somewhat dissenting grassroots contributors harnessed the potential power of the internet (Ward et al. 2005) to hold in balance the Conservative elite, simply using intellectually charged interaction with computerized devices. Although, the extent to which the central party has reorganized its operations in response to the transparent nature of ConservativeHome is not yet apparent.

Another contributor to the forum debate described CPC2012 as the ConservativeHome political conference to which the rest of the Conservative Party had gone along. The comment was made in a debating context in order to provoke thought and reaction, but the speaker's point was significant. Although somewhat tongue-in-cheek, it holds some truth in the sense that when compared with other more traditional Tory affiliated groups and factions, for example The Bow Group and the Tory Reform Group (TRG), ConservativeHome seemed to make the most notable and visible impact at CPC2012.

Speaking at the forum, one ConservativeHome representative claims that ConservativeHome has been part of, and contributed to, an 'internet revolution', which has led to the most radical decentralization of power in modern times. The representative likened this to the significance of the Industrial Revolution, from which time the, now, traditional media began assuming control of the public agenda. ConservativeHome cost $15 per month in set-up cost. The same representative of the blog considers this to be cheap when compared to the launch of a student magazine, which might cost in excess of £400. The point is pronounced further when combined with the blog's daily online readership of 15,000 to 25,000 individuals. The representative believes that there is a huge transfer of power from traditional media channels to the internet, because

DOI: 10.1057/9781137436511.0006

a reader no longer has to wait for media giants like *The Telegraph* to deliver news to them. As in the Tory Bear case above, lesser known and newer brands hold the potential power to stand-up to the larger media institutions using affordable and accessible internet technologies. The ConservativeHome representative predicts that the 'Fordist monopolies' of the big political parties are likely to fall in coming times; and when that occurs, the internet, for example platforms like ConservativeHome, will take their place. Although this prediction seems implausible, it is somewhat in keeping with the ideas of Helen Margetts (2006) who argues the case for a new party ideal type to replace the mass party model in Britain, which she refers to as 'cyber parties'.

In 2005, while Montgomerie worked at *The Daily Telegraph*, ConservativeHome began to transition from an idea into a reality. Respondent Six provides an understanding of the blog's genesis which helps to explain the motives behind its inception:

> ConservativeHome was set up explicitly to campaign on the issue of selection, because they were outraged at the fact that the Conservative Party membership did not really have that much of a choice over who the leader of the Conservative Party was and, as a result, they have always had this campaigning streak – in trying to reform the party and keep the party to a form of Conservatism that the editors and, therefore, by extension, the readers share.

Tim Bale (2010) describes this as Montgomerie acting as a campaign leader for the fight to preserve the party's organizational democracy, especially for those at the grassroots. Therefore, it is plausible to suggest that Montgomerie's longstanding objective for the blog is to siphon some power from the party's centre to the grassroots in order to influence and catalyse reform in the party's organization (how this has manifested itself practically is discussed in Chapter 4). To some extent, Montgomerie's objectives have been realized already in that the central party has become acutely aware of the potential power ConservativeHome could use if it chose to strike-out against the national party. This is notwithstanding the likelihood that the traditional media and opposition parties would significantly capitalize on any major division between the two bodies.

Both respondents One and Six claim that ConservativeHome has been viewed by CCHQ and the Conservative leadership as a 'thorn in the side' of the Conservative Party. Respondent Eight, a Tory MP, believes that in order for ConservativeHome to be credible, unlike the centrally controlled 'Blue Blog', it must remain independent of the central party.

DOI: 10.1057/9781137436511.0006

The respondent believes that ConservativeHome has a significant role to play in party policy and organizational discourse at the grassroots of the party. Even though at times it is a challenge to party unity. The party would be keen to avoid any visible fracture in party unity, because it is a significant factor in the party's electoral successes (Ball 2005). While ConservativeHome maintains an appearance as the face of the Tory grassroots, it is likely to influence party affairs, because it is not helpful to the central elite if it appears to be at odds with its participants at the grassroots.

Respondent Nine, also a Conservative MP, suggests that one of the risks for the party is that in the public sphere the lines can become crossed between ConservativeHome commentary and the official standpoint of the Conservative Party. This suggests that there are concerns within the party about the decentralizing impact of ConservativeHome, and other internet applications, on the central party's ability to control its official communications and key messages. Already blogs have been shown to shift the focus from the central party through offering alternative messages to public discourse (Gibson et al. 2013). Therefore, the primary factor the central party should consider in this case is the blurring of the boundaries between ConservativeHome and the party's organization.

Respondents Eight and Nine provide the 2010 candidate perspective. They agree that the role of ConservativeHome in the run-up to GE2010 was to open greater channels for 'conversation' at the grassroots. For example, they believe it allowed the candidate selection process to be more transparent – rather than a closed affair hidden under the control of CCHQ and Conservative associations. This perspective demonstrates how to some degree the party was being forced by ConservativeHome to loosen its grip on the information and processes that it had traditionally held close to its chest. For example, historically, candidate selection has operated a bit more like the membership process to a closed club than an open contest favouring the types of candidates who have their finger on the electoral pulse (Bale 2010). It is not clear whether the trend in open primary selections since 2009 was the Conservative Party's response to repeated calls by ConservativeHome for greater candidate diversity. It is more likely due to an observable disconnect between CCHQ and local associations. For the central party, there appears to be a loss of faith in some local associations, with rogue tendencies, to make appropriate candidate selections. Therefore, it seems plausible that the party has acted to divert power from associations to local electors using new modes of

DOI: 10.1057/9781137436511.0006

selection. That said, these changes demonstrate the central party operating in a more open, ConservativeHome-style, manner. It is a significant development when one considers that ConservativeHome has led the way in terms of Conservative Party transparency. Again, it is a case of the big guy following the little guy – or at the very least surrendering to contemporary trends, which in itself is a significant outward expression of inner Conservative Party change.

Respondent Five provides a ConservativeHome perspective on this:

> Historically, I suppose sometimes the party would officially not want to have internal [candidate] selection information out in the public domain. But I would say that the nature of the information filter is such that it gets to us anyway. So I think the party before the last election became resigned to the idea that, "Oh well, ConservativeHome will find out anyway – so we might as well just release it." I suppose in that sense the medium of the internet has assisted with creating a bit more openness and transparency about how these things are done.

Therefore, some in the wider party consider the internet to have had some impact on the central party's traditional culture of control and secrecy, believing it to have led to a greater openness, transparency and loosening of centralized processes. Subsequently, ConservativeHome had a good relationship with candidates (Respondent Five). In fact, Respondent Nine suggests that, used as a platform, ConservativeHome is an excellent 'shop window' from which candidates are able to place themselves on display to the wider Conservative Party. Furthermore, she claims that ConservativeHome has an unparalleled position to mobilize activists to key by-election campaigns as it did in the Crewe and Nantwich by-election in 2008. This evidence suggests that the Gibson et al. (2013) claim that blogs are not tools for the mobilization of grassroots activity is perhaps inaccurate in some cases.

Mixed views and mixed messages

Testimonies indicate that the proliferation of Cyber Toryism, like interaction with ConservativeHome, has led to concerns about the impact of non-centralized internet applications on centralized Conservative Party operations. However, they indicate also that there was an understanding within the wider party of the benefits that this new wave of internet-based political innovation was providing for the party organization and its electoral performance prior to GE2010. Respondent Five suggests that ConservativeHome's success was due to its 'unique' and 'niche' role within the party, claiming that the blog has excellent relationships, not only with candidates and grassroots Conservatives, but, also, with the traditional mainstream media.

Respondent Five disagrees with Respondent Eight on the point that the blog is a challenge to party unity and believes that instead it is a democratizing force that has become the primary source of information on the Conservative Party for many individuals inside and outside the party. Respondent Five claims that without ConservativeHome the information would not be available publically on a real time daily basis, explaining that the internet allows for a cross fertilisation of ideas and media. Through ConservativeHome's integration with social media, like Twitter, it means that the conversation is wide reaching and open to anyone (Respondent Five). Therefore, ConservativeHome's unique role in both lobby journalism and as an evolving organelle of contemporary Conservative Party organization, which now functions and impacts in both the on- and off- line worlds, means that the blog is primarily impacting on the evolution of the party's organizational culture from the grassroots up.

It would appear that the range of influences and multivocality that makes up the contemporary Conservative Party has led to ambiguity both inside and outside the party in relation to the dividing lines between official and unofficial spokespersons. This phenomenon, which is considered by some party officials to be a threat to the party's public identity, is emblematic of the diverse globalized world in which we now live. Fragmented channels of communication for which there is greater competition than ever before to reach one's audience have become common place. The following chapter examines how the likes of WebCameron and ConservativeHome have acted as catalysts to internet innovation at the grassroots level, which, in turn, has led to greater fragmentation in the organization of the party's communications in the use of other social media, like Facebook.

DOI: 10.1057/9781137436511.0006

4
Facebook: New Face of Conservative Organization?

Abstract: *Chapter 4 provides an alternative narrative of Cyber Tory activity in that it shifts the spotlight from more elite figures in the party to those at the grassroots. The chapter is rooted in an analysis of the role of Facebook in the party's organizational culture from 2008 onward. It is argued that the culture of Facebook participation evolved naturally through a learning and copying behaviour. The chapter reveals the importance of leadership in this process and identifies specific individuals in the ranks of the younger cohorts of the party as significant influencers of change. The chapter argues that Cyber Tory Facebook participation was characterized by a technologically centred innovation culture that helped dissolve traditional geographical and hierarchical barriers to grassroots activity.*

Keywords: citizen-initiated campaigning; e-participation; Facebook; party organization; political shopping mall; technological innovation

Ridge-Newman, Anthony. *Cameron's Conservatives and the Internet: Change, Culture and Cyber Toryism.* Basingstoke: Palgrave Macmillan, 2014. DOI: 10.1057/9781137436511.0007.

DOI: 10.1057/9781137436511.0007

It is argued in Chapter 2 that Tory web applications and technologies like WebCameron had their developmental stages rooted in an elite ideas culture within Cameron Central. Chapter 3 describes the relationship between the central party and a new media elite which emerged out of the party's grassroots and synthesized with citizen journalism to form a new type of Conservative party organizational subculture. Herein the book expands on those ideas. Chapter 4 presents an alternative narrative that was observed amid the party's grassroots culture in the run-up to GE2010 and beyond. The chapter is rooted in the notion that, from 2006, the publically visible symbols of WebCameron and Ask David, and to some extent the iconic and pioneering electronic grassroots leadership of ConservativeHome, signalled to the wider Conservative party an affirmation of the acceptable use of internet technologies in Tory politics and culture.

The chapter argues that the party developed a grassroots social media culture that evolved and dispersed throughout cohorts within the party in an organic manner. Rachel Gibson (2013) argues that the traditional model of the professional-based party is being challenged by a new form of 'citizen-initiated campaigning', which has been facilitated by the proliferation of new digital technologies in political culture. In that sense, Gibson's work is perhaps revealing a wider trend towards more cyber-based parties in Britain (Margetts 2006). However, unlike Hindman (2009) (Chapter 2), Gibson is arguing that the driving force for such change is rooted in a technologically facilitated devolution of power to the grassroots levels. This is in keeping with the apparent impact that ConservativeHome appears to have had on the Tory Party (Chapter 3), while at the same time the blog has filled a niche in cyberspace and gained itself some form of elite status within the party (Hindman 2009).

This chapter focuses on the role of Facebook and argues that, in the specific case of the Conservative Party, the uses of new media were most effective in campaign and grassroots party organization. There is little evidence in the documentary, ethnographic or interview data presented in this chapter, or elsewhere in the book, to suggest that CCHQ actively encouraged and/or discouraged activists, candidates and/or associations to use email or social media in their local campaigns and/or organization. In fact, the evidence presented in the remainder of the book suggests that in the run-up to GE2010 the central party allowed relative freedom, in keeping with the party's tradition of autonomous constituency

DOI: 10.1057/9781137436511.0007

parties (Ball 1994a), to its individual participants in terms of their use of internet-based political applications. This claim is supported by the Fisher et al. (2011) study of GE2010 campaign activity.

Therefore, the party's inherent culture and tradition of autonomy impacted in the manner in which internet technologies were used. This is significant because the evidence will show a notable heterogeneity that came in the form of experimental and sometimes unilateral approaches taken by specific Tory individuals and associations towards the use of email, social media and websites for political communication. Theorists like Beck (2002) might suggest that this approach is in keeping with globalized social trends of the individualization of people and institutions. If so, then the Tories' approach to local autonomy could be argued to have been ahead of its time. Of course, this is in the context of the Labour Party's historic 'top down' modes of control (Kavanagh 2013).

Prior to GE2010, Conservative Party campaign advisory literature avoided detailing significantly a guide to the best practice for the use of new media and e-campaigning. However, at CPC2012 the party released two documents which indicate a change in approach. The parent publication 'Campaign 2013 Toolkit' (Conservatives 2012b) presents a range of considerations for party activism. The publication makes e-campaigning a special case. The internet is described as 'a powerful and cost-effective campaigning tool' and refers the reader to a separate publication devoted solely to e-campaigning. The 'Online Campaign Guide 2013' (Conservatives 2012c) is a detailed Conservative publication which includes information on using websites, emails and social media as political communication tools and explains jargon and legalities. This signals a post-GE2010 change in the central party's attitude towards new media.

It also suggests that prior to GE2010 the assimilation of new media into Conservative Party culture had not yet been fully realized and/or understood by the central party. Therefore, the party's understanding of the role of new media in its organization appears to have matured since GE2010. It suggests that CCHQ is now beginning to actively engage in the dissemination of educational information on social media and e-campaigning in order to school Conservative participants in 'best practice' for the use of new media in political campaigns and organization, which is similar to the party's approach to TV education in the 1950s.

DOI: 10.1057/9781137436511.0007

Facebook and the Tory grassroots

By 2008, in the run-up to the London Mayoral Election campaign, Facebook had begun being used as an organizational tool for political mobilization in the Conservatives' 'Back Boris' campaign (Respondent One). At that time, the UK was considered to be in the top three Facebook using countries, only behind the US and Canada (Hodgkinson 2008). By 2010, over a third of people in the UK are reported to have joined Facebook (Williamson 2010). Individuals and collectives had begun employing Facebook's social networking capabilities for personal communication and social interaction. These wider trends in Facebook use had begun to be embraced within cohorts at the grassroots of the Conservative Party. According to Respondent One, the turning point for the assimilation of Facebook within the party was when Boris Johnson released a Facebook 'App' for the mayoral campaign. Respondent Six explains how Facebook was used to organize teams of young Conservative activists to mobilize the Conservative vote in the 2008, 2009 and 2010 elections:

> In terms of Conservative Future (CF), the first thing you do is start on Facebook. You would say, "I want to have a campaign day", because my agent in this constituency, or my chairman, or my candidate has said, "I want to have this many activists out." You would talk to them and either get money for refreshments or a lunch or so on. We could then pledge that to our activists and put that on Facebook and you get more people turning up. If you have done your job in getting money from the association for refreshments or lunch, then it is a much easier sell. I organized council by-election campaigns in the local area, in Hampstead and Kilburn, leading up to 2008 and 2009, and we managed to get a lot of local campaigners from UCL [University College London], King's College London [KCL], LSE [London School of Economics] and so on. We did, explicitly, go out saying that we need to get young people involved, because, frankly, the old people are going to campaign in their backyard. The young people are the added bonus that are going to push you across the finishing line. So, we did use Facebook very heavily. I think I set up probably 10 campaign day events for every by-election going, and it is tried and tested – and it seemed to work.

This testimony offers a perspective of a young Conservative Future activist, or CFer, at the sharp end of Conservative Party campaigns in London in the run-up to GE2010. It suggests that younger Conservative activists were more likely than older activists to use Facebook and travel

DOI: 10.1057/9781137436511.0007

outside their local geographical and political boundaries in order to assist campaigns advertised on the social networking site. It reveals that Facebook was used at the local level as an in-house marketing tool in order for key Conservative participants, who were in activist mobilization roles, to sell campaign activities to younger members of the party through a prominent interactive digital medium of their generation. The selling mechanisms used were made more effective when Facebook's direct targeting was combined with the traditional incentives of complimentary refreshment, thus resulting in a quid pro quo campaigning culture at the heart of the Facebook-facilitated activism. In general, these were already politically active individuals who were mobilized into action rather than new activists being generated through the use of social media.

A study by Nils Gustafsson (2012), which used focus groups to assess the role of social networks in political participation in Sweden, had similar findings and concluded that those who were politically inactive were unlikely to be spurred into political activity through social media alone. In the US context, one study of the 2008 presidential elections showed a strong correlation between participation in Facebook groups and offline activism (Conroy et al. 2012). Another study of the same election showed a correlation between political Facebook participation and other forms of civic engagement (Vitak et al. 2011). These findings would suggests therefore that those individuals in the Tory party engaging in political social media activity are likely to be more naturally predisposed to act in that manner and are not necessarily inspired to activism through the advent of new technologies. Moreover, this ties in with the notion that the internet can be a weaker form of interaction in terms of mobilizing new supporters (Margetts 2006).

In the Conservative Party case, Facebook was used as a tool to persuade and mobilize already active Tories to participate in campaigns to which they would not have contributed traditionally, because of geographical barriers. Some aspects of campaign organization migrated to Facebook, the hub of communication that was being used already on a daily basis for general connectivity by younger Tories, which facilitated outcomes of seemingly new social norms (van Dijck 2012).

> People log onto their Facebook every single day and, if you do pester them, then, in effect, they will cave in, which is why, if you do organize 10 campaign day events and only 10 of your activists in your group of 300 friends on Facebook turn-up, that is still 10 activists more than you would have otherwise – and 10

times your 10 campaign days is probably more activists than you will be able to put on the street than the association will itself (Respondent Six).

The use of Facebook in this way helped to make participation in Conservative activism a more fluid and decentralized process (Gibson 2013). At a fringe event during Conservative Party Conference 2012, one activist commented that they believed that social media had brought the party closer together, suggesting that, from a party organization and campaign point of view, it encourages activists to give mutual-aid (campaign support) in other geographical locations. It would suggest that, at the Tory grassroots, there is now, post GE2010, some internal realization of how internet technologies have impacted on the party's organizational culture, coupled with some understanding of its benefits in aspects of political campaigning. The advent of debates like those held at CPC2012, which questioned whether social media has changed the Conservative Party (Chapter 3), is also telling. It shows that, since GE2010, the application of social media in campaigns is being discussed and debated with interest from both inside and outside the party. Therefore, social media has become assimilated into the party's inter- and intra- cultural discursive behaviour.

By 2008, the Conservative Party had begun using online venues as a place to meet the next generation of British Conservatism. Once connected through Facebook, from the comfort of a personal laptop or desktop computer, prominent individuals with skill in using online social networking tools were able to effectively impact on the numbers of activists attending campaign days in the offline world. Respondent Six provides a personal narrative that gives some cultural insight into the discovery and development of Facebook within young Conservative circles around that time.

> I joined Facebook in 2006, when I first went to university. I had never heard of it before the day I signed up for it and there were very few people at university that were on it at that time. One person that used it very effectively was the president of my conservative society at UCL – Richard Jackson. He was very effective, and still is very effective in organizing events and organizing complete campaigns for CF, and others, on Facebook. He did not teach me – he did not sit me down and lecture me on exactly how to do it, but it is good best practice to copy, and it is pretty simple best practice to copy.

This evidence supports one of the key arguments in this book that Facebook, as an organizational tool for the Conservative Party, developed

DOI: 10.1057/9781137436511.0007

organically at the grassroots of the party. In the wider context, this finding can be explained by social movement theory which suggests that the use of social media is an ideal technocultural development for mediating political activism (Lievrouw 2011).

Leah Lievrouw's (2011) theory is helpful in that it strengthens the proposition that (1) Facebook functionality was ideal for the mobilization of party participation at Conservative events and campaigns; and (2) its use grew-out of individual user innovation. In the Tory case, this theoretical explanation is particularly helpful when it is integrated with (1) the empirical observations that organizational leaders – like David Cameron, who first used WebCameron as party leader; Tim Montgomerie, who combined his journalistic status with the internet and his grassroots roles in the party to form a leadership role as the founder of ConservativeHome; and Richard Jackson, who leveraged his skill in Facebook use to enhance his prominent leadership and organizational role within the party's youth movement – acted as Cyber Tory catalysts that promoted the wider use of internet-based media at the party's grassroots; and (2) further theory like that of Edgar Schein on the role of leadership in organizational culture. Schein (2010) states that leaders are hierarchical innovators whose actions direct and drive culture change within the organization to which they belong. Historically, the Conservative Party has maintained a significant organizational hierarchy and deferential culture (Seldon and Ball 1994). Therefore, the leadership has been shown, at specific points in history, to be a significant driving factor contributing to change within the party (Bale 2012). It is, therefore, plausible to suggest that the actions of specific figures within the party helped to speed-up the internet-based culture change witnessed in the party in run-up to GE2010.

In the UCL case, it appears that the leadership actions of a university Conservative society president are significant. The actions were observed and copied by other Conservative participants in other parts of the party organization. In turn, it seems it was this that led to an organic proliferation of the Conservatives' grassroots social media culture from 2008. The ethnographic observations suggest that this learning and copying culture was passed-on from one group or individual to the next. Both the on- and off- line activist behaviour proliferated to provide significant impact for the party's grassroots operations as this form of Cyber Toryism spread. The use of digital technologies by Cameron and Johnson, the face of the Conservative-elite, seem to have signalled to the

DOI: 10.1057/9781137436511.0007

party's grassroots that innovative uses of new media for the party's gain was an appropriate activity in which to engage. Young activists at the Conservative grassroots appear to have responded accordingly within on- and off- line environments, which provided relative freedom for a culture of digital experimentation and innovation that was tempered only by aspects of the party's traditional organizational culture – like the remnants of deference and unity in the party's collective approach to its corporate messages, with the aim of winning GE2010 in mind. It appears that the balance between the individualization of new forms of activism and the collective goal of delivering the corporate message, generally, kept the culture of Cyber Toryism in organizational stasis.

Like many social phenomena which evolve rather than become founded in some act or constitution it is challenging, if not impossible, to outline with any certainty the moment of genesis when Facebook became a significant part of Conservative Party organizational culture. However, one respondent's personal observations offer a perspective on how it may have come about:

> Within Facebook we have had groups and pages and then new groups have come along, and these different architectures are used by different people. I guess there is a bit of an evolutionary aspect to it, in that the people that cannot use Facebook particularly well are kind of nudged aside by other people in the organization in Conservative Future and told to follow the UCL Conservative Society group's structure. They get 400 people a year going to their event, so they obviously know what they're doing. And, I guess, best practice spreads that way, because there's certainly no training days or courses that I've been to on how to use Facebook to get people to campaign (Respondent Six).

Richard Jackson, the 2006–07 UCL Conservative Society president, and more recently a CCHQ press officer, was a close personal friend of, and worked closely with, the 2008–10 CF national chairman, Michael Rock. During that period, the role of the UCL Conservative Society was one of national prominence in the CF movement. UCL Conservatives' close proximity to CCHQ London; the London Mayoral Campaign 2008; and their influential position and relationship with other prominent University of London colleges that had CF societies meant that the society wielded a significant influence on the manner in which CF and its use of internet technologies developed in the run-up to GE2010. Furthermore, the observations and interviews which inform this book show that, prior

DOI: 10.1057/9781137436511.0007

to GE2010, Facebook was largely used by the younger demographic of the Conservative Party. Therefore, it is plausible to suggest that the evolution of widespread use of Facebook within the Conservative Party had its roots in the CF movement, and, perhaps, more specifically in UCL Conservatives.

The UCL approach tends to be in contrast to the 'Oxbridge' universities' Conservative bodies, Oxford University Conservative Association (OUCA) and Cambridge University Conservative Association (CUCA), which operate less as practical campaign resources to the party and more as highbrow social and debate driven policy forums. Both Oxbridge groups have a Facebook presence. CUCA tend to use their open Facebook group to portray the more traditional elite face of British Conservatism. While OUCA have a closed Facebook group and heavily invests in email to reach its organizational objectives.

Richard Jackson's role as a leading figure in the use of Facebook and his position of leadership and interconnectedness between influential networks at CCHQ, the Carlton Club, the Back Boris campaign, the University of London colleges, Cities of London and Westminster Conservative Association and the CF movement meant that he was a key figure on Facebook who utilized his connections to a significant quantity of quality Conservative 'Facebook Friends' interested in activism. Therefore, Jackson's early role in the passive dissemination of the use of Facebook in CF, and subsequently the wider-Conservative Party, may have been an additional significant factor in the development of Facebook being used as a tool for Conservative organization in campaigns in London and, later, nationally.

By the GE2010, the use of Facebook as a political organization tool had become an accepted application for use by many Conservative parliamentary candidates. Respondent Seven, a GE2010 candidate, offers his perspective on the political communication process that a candidate was likely to take when selected for a parliamentary seat.

> One of the first things they [new candidates] did almost straight after they put out the press release or even before they put out the press release was that they set up a Facebook group. You wanted to make sure that you had enough people to join it straight away. You monitored what other political parties were doing in your area on Facebook and you used it as a way of not just signing up existing members. The age profile of constituency association members/officers means that perhaps they were not the most switched on in

terms of e-access, and certainly the use of social media. It was a good way of reaching out...

Respondent Seven is claiming that, in some cases, a Facebook presence for a newly selected candidate took priority over the traditional press release. If so, this would indicate that there has been a significant change in the political communication culture in which some Conservative candidates have operated since the advent of social media. Recent research indicates that similar trends were observable in other British parties in 2010 (Jensen and Anstead 2014).

Through social media, candidates were beginning to be able to take charge to some extent of the dissemination of their own communications through their own channels on Facebook, Twitter, YouTube and/or in using a blog. WebCameron had claimed back from the broadcasters some control and power over output for the central party. Similarly, Facebook shifted some of the power from CCHQ to the local candidate at the party grassroots (Gibson 2013). As Respondent Seven suggests, a further benefit of Facebook was that it allowed Conservatives to group together individuals, who were supporters of that specific candidate's campaign, in digital venues within the social network. These venues were called Facebook 'groups' and/or 'pages'. Facebook, therefore, provided some power to the candidate as an organizational tool for political mobilization of non-geographically bound supporters, which in turn enhanced the candidate's experiences in a local campaign context.

Respondent Seven identifies the significance of having a number of Facebook users showing their support on a candidate's group and/or page. Respondent One goes further in explaining that:

> Facebook is very good from the PR [public relations] perspective: "Great I am more popular than you are." Facebook is literally a popularity contest. The more people that "Like" your page, the more popular you are – as a rule of thumb. In terms of being able to transfer those into votes, I would be very interested to see what the hit rate would be – I don't think it is going to be huge, but it helps.

Respondent One believes that, for some politicians, Facebook was a competitive tool used to demonstrate political popularity. However, he recognizes that if there were a link between Facebook popularity and gaining votes, in 2010, it was tenuous. What is clearer is that some Conservative politicians and their teams believed it to be politically and organizationally advantageous for them to develop a presence and

DOI: 10.1057/9781137436511.0007

audience on Facebook in the run-up to GE2010 – even though there was little explicit direction from the central party on the matter. The extent to which Facebook was used by politicians and political groups, and its efficacy for building an audience and/or campaign support, seems to have varied significantly.

Political Facebook activity

It is important to appreciate the dynamic nature of social media, but especially in the case of Facebook. Facebook's coding architects make regular changes to its functions and how the site operates (see van Dijck 2013: 33). This means that precise retrospective historical research that uses Facebook activity to assess Facebook culture is virtually impossible unless the Facebook data has been permanently recorded in some way at the specific moment of interest. The quality and validity of Facebook data as a record for historical inquiry degrades progressively the further away from the period of interest one begins sampling. This book attempted to restrict such limitations to a minimum in using the relative immediacy of the ethnographic approach.

In preparation for this book, between 1 December 2009 and 31 May 2010, data from a number of political Facebook groups and pages were sampled, observed and recorded. One observation noted that Gordon Brown, Leader of the Labour Party 2007–10, did not have an official political Facebook page. Brown was the only leader of the three main British political parties to not have an official presence on Facebook. On searching the name 'Gordon Brown', a number of pages and groups relating to Brown appeared in the Facebook search results. The themes of these groups were significantly weighted toward calls for him to resign as Prime Minister. As potential substitutes, two alternative leading Labour Party figures, Harriet Harman and David Milliband, were searched. Like Brown, they did not have any official public or political presence on Facebook, other than the official Labour Party page which represented the general central party collective. This would suggest that in 2010 the Labour Party's central operations took a different approach to the Conservatives and Liberal Democrats, whose leadership teams were well represented by official political pages on Facebook. These findings reflect those of Fisher et al. (2011) who found that Labour lagged significantly

DOI: 10.1057/9781137436511.0007

behind the Conservatives and Liberal Democrats in their e-campaigning output.

In observations of a sample of 14 Tory affiliated Facebook groups, the membership figures ranged from 47 for one university CF group to 641 for an established national Tory faction. All of the groups had their privacy settings operating as 'open', which suggests that the information viewable on the groups' pages were deemed by the groups' administrators to be non-sensitive. Analysis of CF university groups showed that there was some correlation between the number of members in a group and the frequency at which the group sent Facebook email messages to their respective members. UCL CF, with its 250 members, sent an average of three emails per month. This is compared to KCL CF's 154 members receiving 0.83 emails per month; and Aberystwyth CF's 74 members receiving 0.17 emails per month. In these cases, it would appear that the greater the membership of a group, the more frequently Facebook messages were sent members by the group's administrators to group members. A similar trend is observable when comparing the association-based CF groups. With 384 members, Cities of London and Westminster CF had the largest membership of any of the CF groups and sent an average of 2.67 Facebook email messages per month. Richmond Park CF, with its 173 members, sent 0.83 emails per month. Aberconwy CF, with its 47 members, sent 0.50 per month.

The trend continues further when comparing the candidate groups. At 261, Nigel Huddleson had the largest number of members out of the candidates sampled and sent 0.67 emails per month. Michelle Tempest had the lowest, with 161 members, and sent 0.33 emails per month. A similar trend is shown in the Tory faction groups. The Bow Group, with its 609 members, sent on average one email per month. Progressive Conservatives had 423 members and per month sent 0.67 emails. The TRG had 394 and sent 0.33 emails per month. Amid the trend there were outliers. Unlike the other Tory faction groups, Conservative Way Forward (CWF) sent no Facebook emails to its 641 members in the 26 week period. But it did use traditional email eight times to communicate with its members, which it addressed as 'Colleagues'. This compares to no group-sent emails being observed for The Bow Group. Therefore, it would appear that while some Tory groups chose to use Facebook email to communicate with its participants, others preferred the use of more traditional email methods.

DOI: 10.1057/9781137436511.0007

Furthermore, CWF had a lower Wall posting activity than the other Tory affiliate groups, whereas the Progressive Conservatives were significantly more active in the number of items posted to their Wall. Therefore, a diverse range of activity was observed between the groups. However, there were some observable trends in terms of Facebook Wall activity. The Facebook groups with the largest memberships of the university CFs (UCL), association-based CFs (Cities of London and Westminster CF) and candidates (Nigel Huddleson) categories showed the highest activity in terms of administrator Wall posts. These same groups showed significantly higher event posting activity. This would suggest that the most popular Conservative Facebook groups were those which were more social and more active both on and off Facebook. Moreover, it would suggest that there was a direct and mutually dependent relationship between on- and off- line political activity; and that, in the run-up to GE2010, having a strong presence in both the on- and off-line worlds assisted the growth of Conservative support both on and off Facebook. This notion is in keeping with the findings that support the Social Technology Acceptance Model (Paris et al. 2010) in which there is shown to be a relationship between on- and off- line Facebook event participation.

In the run-up to GE2010, the functionality of Facebook groups leant toward uses in Conservative Party organization. Facebook group functions enabled administrators to organize, promote and communicate easily in both social and campaign event contexts. The captured audience within a Facebook group received information from digital event notifications and Facebook email messages when group administrators published event information. This meant that event organization was targeted, simpler and faster. Its application on Facebook was also more dynamic than other electronic alternatives like email technologies. The administrator could monitor guest/RSVP lists in real time as members actively responded with the click of a Facebook button to invitations via their Facebook profile. Therefore, Facebook groups played a significant role in changing and enhancing the organizational culture of those groups within the Conservative Party which actively used Facebook groups to develop their offline sociality and campaign operations.

However, it is important to note that in 2010 Facebook pages functioned in a different way to Facebook groups. One significant difference is that, unlike group members, there was no way of emailing Facebook page 'Fans' collectively. Pages were globally public within Facebook and communication activity was centred primarily on the page's Facebook Wall. Individuals with a Facebook profile could be invited to, or choose independently to, 'Like' a page and therefore become a fan of that page. Pages were used to promote both individuals, for example David Cameron as a politician, as well as collectives, for example the Conservative Party as a political movement.

Political shopping mall

Jensen and Anstead (2014) suggest that candidates with lower profiles might invest time in social media in order to compensate for their lack of presence within traditional media output. In that sense, the notion that Facebook pages act like shop windows for both individuals and collectives is a view held particularly within the Conservative Party (Respondent One; Respondent Nine; Lee 2014). Websites also act like shop fronts, except they are accessed by a loosely organized global audience. Facebook is populated by a body of individuals who make a choice to develop a personal presence within its password protected, and, therefore, more tightly organized, semi-closed, online community. Through their interests, Facebook users make additional choices to congregate online as members of Facebook groups or as supporters of a cause, an individual and/or a collective through expressing an electronic thumbs-up. Therefore, in the case of politics, Facebook pages act like shop fronts situated in a distinct online community – rather like an online political mall in which the political consumer, first, makes the decision to go shop at the mall; and, second, tour and browse some political shop fronts before publically buying into political brands by becoming a 'Facebook Fan'. It is like carrying a branded carrier bag and displaying to all at the mall that you have bought in to a particular political product. In the political popularity contest that is facilitated by Facebook pages, the idea is to encourage as many users to carry your political carrier bag as possible. According to research by Southern and Ward (2011), the Conservatives, Labour and the Liberal Democrats had fairly equal levels of grassroots Facebook use in GE2010. Therefore, these findings would suggest that Facebook was beginning to impact across the political spectrum, thus allowing a greater diversity of politics to be presented to the Facebook user in the political shopping mall. However, as described above, the same cannot be claimed about access to the Labour Party elite,

DOI: 10.1057/9781137436511.0007

which was largely non-existent on Facebook in an official capacity. This would suggest that in terms of cyber political culture in 2010, there was a division between grassroots and elite cultures in certain British parties.

A simple comparison of Facebook page 'Likes' shows how the scale of interactive activity differed significantly from Facebook groups to Facebook pages. The Facebook pages yielded a significantly greater level of interaction by non-administrators than Facebook groups. Therefore, where Facebook groups encouraged a shift in Conservative organizational culture both on- and off- line, it would appear that Facebook pages attracted significantly greater online activity in terms of the volume and frequency of interactions. The number of 'Likes' or 'Fans' displayed on a political Facebook page in 2010 could differ considerably. Lower profile politicians like David Jones, Kwasi Kwarteng and Robin Walker had quantities of page likes ranging from Walker's 125 fans to Jones's 346 fans. The higher profile politicians including Boris Johnson, David Cameron and Nick Clegg each had tens of thousands of fans, from Johnson's 49,733 to Clegg's 73,084. In contrast to Southern and Ward's generalized study, the specific Facebook cases observed here indicate some differences between the ways in which participants interacted with individual political party Facebook pages. Comparing the Clegg and Lib Dems pages with the Labour and Conservative pages shows that the campaign events activity postings was low for Labour and the Conservatives, but higher for the Liberal Democrats. Therefore, the Lib Dems were more inclined to advertise their campaign pursuits on their main Facebook page than the two other main parties. This suggests that Labour and the Conservatives considered that their campaigns had strategically more to lose in doing so than the Liberal Democrats.

However, there was greater posting activity by the Conservative leadership than the Liberal Democrat leadership in terms of photographs and videos. There were 56 photos and 34 videos posted by page administrators to Cameron's page compared to 14 photos and 32 videos posted by page administrators to Clegg's page. The Conservatives' page, which had the most number of fans at 111,540, had also the greatest activity in posting 78 photos and 71 videos. The Labour page, which had the lowest number of fans of the three main parties at 61,485, posted 0 photos and 21 videos. The Liberal Democrats, with 91,878 fans, posted 15 videos and 53 photos. These figures demonstrate how in the run-up to GE2010, the Conservative Party was the most popular party on Facebook, even though Clegg was the most popular political leader of the three main

DOI: 10.1057/9781137436511.0007

British parties. The Labour Party was the least popular party and it had significantly lower activity in terms of integrating its publicity and media with its Facebook pages.

The Conservative Party was the most prolific in terms of updating its centralized Facebook pages and integrating them with visual media. There was also greater activity by both the Conservatives and the Lib Dems in terms of posting links to external sites. Therefore, the two parties used Facebook to actively promote their other online interests like blog postings and links to central party website pages with greater intensity than Labour. The average likes per Wall post is a good indicator of public interaction by non-administrators. The pages of the higher profile politicians and the main political parties ranged from 78.31 average likes per administrator posting for the Liberal Democrat page to 148.06 average likes per Conservative page administrator posting. Overall, the Conservatives were consistently the party with the greater administrator and non-administrator public interactions on their Facebook pages. Therefore, it is plausible to suggest that the Conservative Party held the greatest intensity of activity on Facebook in 2010. The Liberal Democrats were not far behind the Conservatives in the levels of interaction with the new medium. However, Labour were distinctly less engaged with the use of Facebook's public pages.

Dissolving barriers

Facebook was used effectively by the Conservatives to encourage participation within the party organization by new and more established party supporters. Facebook groups were used as a tool to organize events and campaigns. Facebook pages became for some candidates a political shop front. It acted as a display window in which the administrator could furnish the page Wall with visual multimedia including text, photographs, videos and hyperlinks. These could be used to sell the candidate or political party to a new market of political consumers. The candidate or collective group sold their causes to captured audiences in new ways using the organizational ease of a low cost digital network community. This was quite a significant and symbolic advance on the political communication culture of previous general elections (Lee 2014).

In using Facebook, candidates and activists at all levels of the Conservative Party had access to, for the first time, a medium which held

DOI: 10.1057/9781137436511.0007

the potential for relatively unknown politicians to develop an audience and demonstrate their popularity in a publically viewable manner (Jensen and Anstead 2014). Ordinary candidates and activists had the potential to challenge the traditional party hierarchy in having the opportunity to engage with a medium in which both the grassroots participant and party leader had access to the same platform. Therefore, Facebook's role in the Conservative Party contributed to a cultural change in the daily practice and use of political technologies in an array of areas in the party's organization and for a significant number of individuals.

By 2010, Facebook was acting as a venue that brought together likeminded individuals in locales within cyberspace which did not discriminate in terms of spatial limitations and geographical boundaries. Facebook brought closer together than ever before candidates and potential activists from across the country insofar that, with the immediacy and localization of Facebook through the internet, geographical boundaries were viewed as being much less limiting. In doing so, it removed the reliance of candidates and activists on the traditional party structure which had been long dominated by CCHQ and the national party organization (Gibson 2013). It would seem that this use of Facebook had begun to dissolve the traditional and historic barriers and boundaries for candidates and activists at the grassroots in terms of political communication, which, since the 1960s, had been largely dominated by television and the gradual centralization of political profile and output. Therefore, Facebook seemed to further empower for the Conservatives a dynamic grassroots communication culture, which is in line with the theories about technological impacts on internal democracy that is cited to indicate to a more networked campaign culture in the UK political context (Gibson and Ward 2012). As the oral testimonies indicate, Facebook appears to have allowed and facilitated easier organization of offline sociality; otherwise unknown participants to develop a profile through which their voice was more readily heard; and the opportunity to promote messages outside of those dominated by the traditional centralized control.

Most significantly, Facebook, as an internet application, contributed to a technology centred innovation culture at the grassroots, which evolved and spread through a learning, adapting and copying behaviour by Conservatives who used the medium early on in the 2005–10 election cycle. However, it is important to be cautious about generalizing Facebook behaviour in this context, because of the demographic and leadership trends that played significant roles in the events that led to

DOI: 10.1057/9781137436511.0007

its evolution as a political and organizational tool within Tory culture. A range of respondents have supported one of the key observations that, in the run-up to GE2010, the majority of Conservative-minded individuals interacting with politician pages on Facebook were representative of the younger wing of the party (respondents One; Three; Six; and Nine). The testimonies have provided narratives which are useful in understanding how unwitting leadership within prominent CF groups led to the passing down of Facebook best practice in an organic manner through observation and learning.

It was through this behaviour that a new and distinctive internet based cyber culture within CF began to proliferate. The culture of Cyber Toryism in turn led to a loosening of the control that the central party had over party organization and its communication and campaign operations similar to that described by Gibson (2013). Therefore, it seems plausible to suggest that both Hindman's (2009) elite niche perspective and Gibson's (2013) decentralization perspective are helpful in explaining specific phenomena observed in the organizational culture of the Conservative Party in the run-up to GE2010. However, it is pertinent to note that both perspectives are not mutually exclusive in the Cyber Tory case. It would appear that Cyber Toryism was a complex soup of e-interactions which led to a general redistribution of power and reorganization of networks at both grassroots and elite levels and across Conservative individuals, groups, institutions and systems.

DOI: 10.1057/9781137436511.0007

5
In the Net: Joining Cameron's Conservatives Online

Abstract: *Chapter 5 takes a look at the journey of becoming a Conservative Party member from the participant's perspective. The chapter identifies how the party's online processes were out of sync with its traditional membership structure and that, while the party was in transition in the run-up to General Election 2010, the party seemed to lose some active engagement potential from its online membership because the party was ill-equipped to convert weaker forms of online membership in to stronger forms of face-to-face participation. The chapter provides the narrative of an ethnographer journeying from an online political neophyte to a fully initiated and active member of Cameron's Conservatives.*

Keywords: Conservative associations; Conservatives.com; ethnography; General Election 2010; online membership; party membership

Ridge-Newman, Anthony. *Cameron's Conservatives and the Internet: Change, Culture and Cyber Toryism.* Basingstoke: Palgrave Macmillan, 2014. DOI: 10.1057/9781137436511.0008.

DOI: 10.1057/9781137436511.0008

In previous chapters, the evidence for Cyber Toryism has been primarily placed within the national Conservative Party context. The remaining case study chapters continue to build the case for Cyber Toryism, but in the more local and geographically specific contexts. These chapters have more of an ethnographic sensibility in that there is a greater focus on the use of the first-hand insider perspective of the author. The approach to these chapters is influenced by 'autoethnography' in which the autobiographical narrative of the researcher's memoirs are subjectively embraced in addition to the observation of others (Ellis and Bochner 2000; Jones 2005; Anderson 2006; Chang 2008; Siddique 2011). Therefore, the use of 'I' is more prevalent in the remaining chapters in order to help the reader distinguish between the first-hand experience of the author and the testimonies of others. Chapters 6 and 7 form ethnographies that were conducted while I was an active participant in the field with Cameron's Conservatives, at the local grassroots level in Surrey and Anglesey.

Chapter 6 includes some key findings based on my first-hand observations as a Conservative participant in the cohorts and clusters of the Runnymede, Weybridge and Spelthorne Conservative Group (RWSCG) and in CF. Chapter 7 takes a similar approach to examining my experience while in the field with Anglesey Conservatives, during my time as the Conservative PPC for Ynys Môn | Anglesey. However, before presenting the two geographically-tethered case studies, this chapter provides, firstly, a narrative of my experience of using the internet to join Cameron's Conservatives; and, secondly, a comparison of the cultural factors relating to the investigation of Tory organization in the two different geographical locations.

Becoming an initiated Tory member

Before becoming actively involved in the Conservative Party, I joined the party membership for a number of consecutive years using the party's online sign-up function on its website at Conservatives.com. On each occasion, I was sent a national membership card attached to a welcome letter. In 2006, the membership welcome letter gave me a warm welcome to Team Cameron. It indicated that I had become both a national and local party member and that my local Conservative association would get in touch with me. I was instructed to go to the party's central website and enter my postcode if I wanted more information on my local

DOI: 10.1057/9781137436511.0008

association. The letter also invited me to consider the practical ways in which I might help support the Conservative Party reach its electoral aims, like, for example, delivering leaflets (Maude 2006).

At that time, I resided in both Devon and Worcestershire and, although my membership had declared me as being a member of 'Cameron's team', I was not directly and personally contacted by any member of a local Conservative association. Therefore, my affiliation with the Conservative Party remained tenuous in that, for a number of years, I had no physical face-to-face contact with any individual known to me to be involved in Conservative Party activism. I was on the periphery of the party organization.

Between September 2007 and August 2008, I was a resident of the Spelthorne Borough. Subsequently, I updated the party with my address details using Conservatives.com. However, as a Spelthorne Borough resident, I did not appear to be reassigned with any immediacy to the Spelthorne association by CCHQ. The local association had no record of my party membership and were not informed by CCHQ in good time that I had moved to Spelthorne. The convenience of using the internet for membership subscription meant there was a degree of trade-off in which my expectations of being a party member were not being met.

The view of one Conservative association chairman offers a deeper perspective on the impact of changing trends and uses of technologies within the membership organization of the party.

> I think one of the many reasons why the party has had a membership crisis over the last 10 to 15 years is that the whole nature of politics and membership, and people's attitudes to parties, have changed. One of the things that used to happen before membership was computerized was that the branches had responsibility for, and ownership of membership, so that every October they would go out and knock on the doors. They knew the people. They said, "Hello Doris, how's the dog, are you going to renew?" This kept the branches in a healthy state of mind. Once the membership was computerized, we moved to a situation where the association would send out standard mail merge letters, that's an impersonal form of communication, and it may not be followed up in a timely way. That probably was the only way to do it because this army of volunteers that we had in the 90s was gradually disappearing, so I don't think there was any option about that change (Respondent Three).

This suggests that these changes in the party's approach to organizing party membership were inevitable and that the party has followed a wider trend due to changing attitudes towards political parties. It indicates that

DOI: 10.1057/9781137436511.0008

individuals ranging from central party elites to local association elites hold the view that this shift in the party's culture, away from face-to-face relationships between party and member, are 'impersonal', but a necessary organizational evolution.

I experienced this first hand. Two years after becoming a 'paid-up' member of Cameron's Conservatives, before this ethnography began in October 2008, I made the personal decision to become more actively involved in Conservative Party activism and events. Therefore, in October 2007, I took the step to contact by email an officer of Spelthorne Conservatives. However, the process of being recognized as a paid-up Conservative member was challenging. In reply to an email received from the Spelthorne officer, I wrote:

> I am a little confused by the process of joining Spelthorne Conservatives. I recently paid my party membership online, which will take me through to November 2008. I am not sure what you mean by my joining in January. Will I have to pay another subscription then? Should I be returning the membership form that you attached even though I am already a party member? (Ridge-Newman 2007).

As a Conservative member who had not yet been initiated in the particulars of active Conservative Party membership, I did not understand at that time that the party was running two different membership systems that were out-of-sync. Online membership was renewable 12 months from the date it was purchased. As a national online membership, it was less directly associated with the local Conservative Party and was, therefore, a weaker link to the party (Margetts 2006). Although the system did automatically assign national members to their local association, it was for officers of that association to act in terms of making contact with the member. However, if a member paid their membership subscription directly to an association, there was greater likelihood that the association would make some contact with that member. Association membership generally ran on an annual basis and was collected in January every year, but, as an internet member who had not yet been involved with an association directly, I did not realize it.

It would seem that at that time the party's online membership operations had not been smoothly integrated within the wider organization of party membership. Therefore, relatively new and peripatetic members, especially those who had joined the party online only, and had no previous formal contact with a Conservative association, were at a significant disadvantage if they were interested in becoming actively

DOI: 10.1057/9781137436511.0008

involved in the party. Internet membership appeared, on the surface, to be representative of Cameron's Conservatives' approach – that all were 'warmly welcome' in the Conservative Party. However, the reality was that becoming involved with some local Conservative associations was more in-keeping with joining a closed social club (Respondent Seven). Already initiated individuals with prior experience of the cultural workings of the Conservative association were at an advantage for active entry into any newly approached association. As an internet member, I had to actively seek out ways of making contact with the party at the grassroots. Therefore, although the party's website was a convenient tool for attracting new members and donations to the Conservative Party, it would seem that its role had not been expanded within the organizational processes to a point at which it fully facilitated active engagement for online members. Dennis Kavanagh (1972) describes political socialization as the process through which a citizen comes to understand their own socio-political development and identity within the democratic system. I would suggest that this applies also here, in the subcultural context, whereby individual Conservative participants learn the cultural and hierarchical ways of Conservative Party organization and networks.

Now that I have been, more or less, fully initiated as a party member, in retrospect, I would suggest that the party's lack of integration of its two types of membership organization in the early run-up to GE2010 was in certain cases a limitation to it integrating new members and potential activists into its workforce at the grassroots. Therefore, this suggests that CCHQ's strategy and interest, in terms of membership subscriptions, were focused on raising central funds via the internet, thus leaving the associations to their own initiatives in terms of nurturing a local support base. The extent to which this was achieved was largely dependent on the approach and leadership style of each association. Unlike in 2006, in 2010, the national membership welcome letter (Pickles 2010) lacked any indication of, firstly, whether an association would be in touch with the member; and, secondly, how the member could research the location of their association or where the information could be found.

Although the central party and many Conservative associations had their own websites in the run-up to GE2010, it appears that there was the assumption that potential supporters of the party would intrinsically understand the complex organizational intricacies of British Conservatism; and, therefore, that the individuals interested in joining the party via the internet in the pursuit of grassroots activism would eventually find their

DOI: 10.1057/9781137436511.0008

own way to active engagement from a generic online national party application process. In a society that had become increasingly fragmented by technologies, in this case it would seem that it had become incumbent on the individual political neophyte to journey their way through a process of discovery in order to make first contact with their local association. This indicates that the Conservative Party's approach to internet subscriptions fits a changing political culture away from one that is rooted in traditional mass-based characteristics. Davis (2010) argues that the role of new technologies in British political culture may have impacted to create a more bloated centralized elite, which in the Conservative context could be described to contain the likes of Tim Montgomerie and Richard Jackson (chapters 3 and 4). This could explain why, on the one hand, Cyber Toryism appears to have democratized party culture, facilitating new elites at the grassroots, and, on the other, disenfranchised some cohorts of participants at the periphery (chapters 2, 3 and 4).

Before I became initiated into the Conservative-fold, I was held in suspended animation at the party's periphery while I interacted regularly via email with an association officer of Spelthorne Conservatives, which eventually developed from a more distant exchange of electronic texts, in the form of emails, to a more personable mobile telephone conversation in which we began to develop some trust and rapport (Hesse-Biber and Griffin 2013). It is important to note that, at that time, I was an unknown entity for the Conservative Party, both nationally and in the individual locations around Britain in which I had lived. In this case, the Spelthorne Conservatives officer was cautious about embracing a new aspirant activist and politician who, at least seemed to have, appeared from nowhere into an email inbox. This is an example of how email and other electronic exchanges can be an inferior substitute for initial face-to-face interaction (Tanis and Postmes 2007) for a Conservative participant with the aim of becoming engaged, to some degree, in party activism. The manner in which this newer form of electronic communication both facilitated a connection and yet acted as a barrier to deeper involvement also correlates with the general trends to which Respondent Three testifies above.

In the New Year of 2008, the Spelthorne officer began advising me of how I could engage in the association on matters of activism and how I could develop my personal political aspirations. Once trust and rapport had been established mutually through telephone conversations and face-to-face contact, my progress in the process of developing my standing within the party accelerated considerably. Later, I was interviewed by the

DOI: 10.1057/9781137436511.0008

association and placed on the Surrey County Council Candidates List, 30 April 2008. However, I made the personal decision not to stand for selection at the Conservative branch level of Spelthorne Conservatives.

In August 2008, I became a resident in the Runnymede and Weybridge Constituency and subsequently joined Runnymede and Weybridge Conservative Association (RWCA). Because both the RWCA and Spelthorne associations were operated in the same building, using the same staff and technology, I had made already the inside connections with the relevant administrators who subsequently made easier my transfer of membership to RWCA. I was by that point an initiated member of the Conservative association class and, therefore, had a greater understanding of the culture in which it operated. Therefore, my transition of integration to the somewhat different culture of RWCA seemed smoother and more welcoming. However, it seems significant to note that personalities were a factor in that transition. I felt more naturally aligned with the approach and culture of the RWCA.

From my perspective, RWCA appeared calmer, less eccentric and more established. In Spelthorne, I felt like an outsider and responsible for integrating myself. The seemingly heterogeneous nature of Conservative associations, even those within the same grouping, means that the most local association is not necessarily the best fit for the individual. In RWCA, perhaps by the virtue of the fact that, through my interaction with Spelthorne, I had become a known entity to the local party, I felt more at home with the officers running the association and within my local branch of Virginia Water and Thorpe – as I did when I moved to Anglesey. The internet being without geographical bias seems a resource that the party could harness. However, how it overcomes the initial barriers of online membership seems, even in 2014, yet to be resolved.

Geographic cases: Anglesey and Surrey

When two associations within the same Conservative group can demonstrate significant cultural heterogeneity, then it will come as no surprise that associations separated by large geographical differences can be rather different too. Comparing Conservative culture in the two major geographical case studies of Anglesey and Surrey reveals some significant differences. Surrey is largely a location that serves as a commuter belt for employment in London and the surrounding areas.

DOI: 10.1057/9781137436511.0008

Anglesey is, traditionally, an agricultural island community, which, due to its remote location, has struggled to maintain buoyant alternative industries and employment for the islanders. Surrey is known as one of the most affluent counties in the UK and Anglesey one of the poorest (Jones-Evans 2009; Salman 2012). The RWSCG had significantly greater pooled resources and facilities than Anglesey Conservatives in terms of membership figures; fundraising potential; the number of hours worked by employed staff; the number of volunteers; value, condition and size of property assets; and the age of and access to CMCs. The RWSCG had a relatively large and inviting two storey, four bedroom, detached property, which housed their joint association operations and was valued at £375,000 (shared equally between the two associations) in December 2010 (RWCA 2010).

Until 2009, the Anglesey Association office had been located for a number of months in a domestic village residence belonging to one of the association officers. However, by 2010, a property that was owned by Anglesey Conservatives was opened in Llangefni town centre for use by the Conservative PPC. According to an informant on the association executive, the property was valued at approximately £50,000 in 2010. The 'two-up-two-down' mid-terrace was in poor decorative and structural order and had a number of visible areas of damp. In comparison with RWSCG, the office was poorly maintained and uninviting. The computer and printing facilities of Anglesey Conservatives were aging and significantly inferior to those shared among the RWSCG. The upper floor of the Anglesey property was rented to a live-in tenant. The public entrance door opened on to the Llangefni high street. The office users and the domestic tenant shared the main access way. The one-room, ground floor Anglesey office was used for administration; a campaign headquarters; and the advertised location for surgeries in which constituents were invited to meet with the Conservative PPC. It appears that my experience of the Anglesey association is similar to Alexander Smith's view of the Conservatives' local office on Castle Street, Dumfries, c. 2003, which he reports as suffering 'from a lack of modern equipment and resources' (2011: 91). In contrast, the RWSCG shared both a part time agent and a part time administrator, in addition to the group's two MPs' parliamentary staff.

In 2009, in terms of membership and voluntary support, individual Surrey Conservative associations had significantly greater numbers of paid-up members than Anglesey. RWCA had 678 registered members, and Spelthorne 403. South West Surrey was the Surrey Conservative

DOI: 10.1057/9781137436511.0008

association with the largest membership. Its 1,382 registered members represented 13.5 per cent of the total 10,223 Conservative Party members in Surrey (RWCA 2010). Out of 11 Surrey associations, Spelthorne was ranked lowest in terms of membership numbers, with 4 per cent of the total Surrey membership. RWCA ranked two places higher with 6.7 per cent. According to an informant with access to Anglesey Conservatives' membership data, Anglesey's association had 134 paying members. This indicator alone demonstrates how the lowest ranking Surrey association in terms of membership had considerably more members than the Anglesey association. Moreover, the Anglesey membership total was less than a tenth of Surrey's highest ranking association. This illustrates the wide ranging scale of membership figures from the higher performing associations like Surrey to the lower-performing associations like Anglesey. The membership figures in a Conservative association indicate the fundraising potential and therefore the financial strength of an association.

Stuart Ball (1994a) relates the strength and weakness, in other words the degree of 'autonomy' of associations from the central party, to the financial independence of the local party. He suggests that these factors can have a significant impact on the culture in which an association operates and the nature of its relationship with CCHQ. However, it is important to note that in 2010, the financial health of an association was not necessarily an indication of how much support a candidate had in terms of grassroots workers during a campaign. Anglesey was a relatively poor Conservative association, but the campaign drew activists on action days with numbers up to 27 participants. This was reasonably high in comparison with other local campaigns, such as the Conservative target seat of Aberconwy.

Each individual campaign illustrated in the following geographical case studies is considered to be unique. They are constituted by a range of variables which include, but are not limited to, the candidate; the geographical location; the election type; the individuals within the campaign team and association; and the point in history in which the campaign is being fought – which in turn influences the technologies used, the pertinent policy issues of the campaign and the economic backdrop for funding the campaign. Whiteley et al. (2002) describe influences on campaign outcomes as 'controls'. Their generic controls include constituency 'social characteristics', 'percentage of owner-occupiers', 'incumbency', 'marginality', and national and local socio-economic and political events. These factors have a dynamic relationship with the macroculture of the national Conservative Party organization; and also the sub- and micro- cultures of regional and local specifics both inside and outside the Conservative Party. For example, the conditions under which my personal campaigns in Virginia Water and Anglesey were fought in some respects exhibited crossover and were in other ways dissimilar.

Whiteley et al. (2002) describe local Tory organization as being fragmented because each association has been historically autonomous. Therefore, naturally, being a Conservative local government candidate in Surrey is different in many respects to being a Conservative Parliamentary Candidate in Anglesey. This is especially the case in terms of public profile; the local impact of the candidate; the daily role of the candidate; and the candidate's relationship in relation to the organizational and hierarchical aspects of the Conservative Party. Therefore, one could argue that the two examples do not make a logical or appropriate comparison. That said, in terms of the cultural basics of campaigning in the two locations, in many respects, the approaches and traditions of the two associations were found to be remarkably similar. There were both cultural and organizational characteristics of my campaign as the Conservative PPC in Anglesey which felt very familiar because of my prior experiences campaigning as a Conservative candidate for a seat on Runnymede Borough Council – for example, canvassing door-to-door and delivering leaflets in groups of Conservative participants, and the common Tory rituals and customs, like wearing blue rosettes and using certain textual and linguistic codes to denote opposition parties.

DOI: 10.1057/9781137436511.0008

6

Surrey Conservatives and the Internet

Abstract: *Chapter 6 details the first-hand observation of the participation of the researcher and others within the Runnymede, Weybridge and Spelthorne Conservative Group context in Surrey. The chapter features an analysis of the role of internet technologies in a local council by-election campaign in 2009 from the candidate's perspective. It is argued that there was both an age and digital divide observable amid cohorts in the Surrey Conservatives and that trust and rapport building were central to dissolving cyber-based barriers to deeper engagement within local Conservative associations. It is found that the internet facilitated new network interactions that made association and campaign organization a looser and more fluid experience, which ultimately led to richer and more diverse campaign-based and social interactions in the offline world.*

Keywords: Conservative associations; e-participation; European Election 2009; Runnymede; social media; Surrey Conservatives

Ridge-Newman, Anthony. *Cameron's Conservatives and the Internet: Change, Culture and Cyber Toryism.* Basingstoke: Palgrave Macmillan, 2014. DOI: 10.1057/9781137436511.0009.

Building on the concepts, narratives and technological phenomena introduced in previous chapters, the Surrey case study is comprised of four sections that place under the microscope the role of the internet in the local grassroots context of the RWSCG and the Virginia Water Ward by-election in 2009. While the previous chapters focused on specific technologies, this and the following chapter take a more holistic look at how a range of internet technologies began to integrate, in more practical terms, into the life of the Conservative associations. The geographical case study chapters take a particular interest in the behaviours, cultures and practices of their respective associations in the run-up to GE2010.

Runnymede, Spelthorne and Weybridge Tories

In early 2008, I was an under 30s member of Spelthorne Conservative Association. Subsequently, via email, I was put in touch with, and invited by, another young/er member of Spelthorne Conservatives to attend a CF event in London on 5 March. The event by its very nature was social and hosted in a public London bar. It quickly became apparent to me that the central interest for most of the individuals at the event was to network socially within a Tory cohort. Therefore, I participated in the customs that I was observing. I met a large number of CFers, many of whom were young professionals who freely disseminated their business cards to individuals with whom they had developed a rapport. By the end of the evening, I had collected 12 business cards. Likewise, I reciprocated by giving out a personal card that I had had professionally printed with my name and email address.

A number of individuals suggested that we should 'find each other on Facebook'. Seemingly, this phrase was used as a social cue in order to indicate a mutual interest in connecting with other young Conservatives. Therefore, the advent of Facebook, used by the CFers as a networking tool, had begun to facilitate social interactions in both off- and on- line Tory social gatherings. Alexander Smith (2011) describes his first networking interactions in Dumfries c. 2003, before the advent of Facebook, as a challenge, because the most enthusiastic individuals he encountered were those who created barriers to him connecting with other local Conservatives. In contrast, I found connecting with CF networks through the use of Facebook comparatively fruitful and immediate, which demonstrates the impact of Facebook in acting to facilitate

DOI: 10.1057/9781137436511.0009

some divergence in the manner in which Tories were making internal network connections.

I had been a member of the social networking website 'Facebook' (Facebook.com) since 2005. Therefore, on my return home to Spelthorne from the London event, I checked my Facebook account online, via a personal laptop computer. In the two hours it took for me to travel home, I had received nine Facebook 'Friend Requests' from individuals I had met at the event. I subsequently reciprocated by 'accepting their friend requests' via my personal computer, and sending friend requests to a number of other individuals whom I had met at the event. Unlike Smith's experience in Scotland, these Tories seemed, perhaps unsurprisingly, abundant in South East England.

Following my first networking event, I was invited to a number of Facebook groups that were used by young Conservatives. In the following two months, the number of CF Facebook events and CF friend requests that I gained grew significantly. Through that process, I made an influential contact in Surrey CF. The Surrey CF Area Chairman, who later defected to UKIP and began writing a blog for *The Daily Telegraph*, had been appointed to oversee the organization of the CF branches in the county. In mid-2008, I was invited to join the Surrey CF Area committee as the branch development officer. I accepted the position and became involved in a number of Surrey CF events. However, my role in Surrey CF did not mature until I moved residence and switched my membership to RWCA in August 2008. This indicates how, although Facebook was a useful tool for the organizational aspects of CF, my growth as a Conservative participant was reliant on having roots imbedded within a fertile Conservative association.

I had already developed face-to-face relationships with the chairmen of both Runnymede and Weybridge and Spelthorne Conservative associations by the time I worked more closely with Surrey CF. I decided to develop a new CF branch to serve the geographical area covered by the RWSCG. Therefore, I announced the proposal of the branch to the chairmen in an email and was subsequently invited to the home of the RWCA Chairman in order to discuss founding a CF branch. Afterwards, via email, I proceeded to announce the following to a number of Surrey Conservative officials:

> When I met with [the RWCA Chairman] at his home on 3 September 2008, I bumped into [the Surrey Conservatives Chairman] who said he would like to connect with Surrey Conservative Future. As Surrey CF Branch Development

DOI: 10.1057/9781137436511.0009

Officer, I said I would put [the Surrey Conservatives Chairman] in touch with [the Surrey CF Chairman]. [The Surrey CF Chairman] has suggested meeting before ... return[ing] to university. I have also discussed branch development in Virginia Water and Thorpe with [the RWCA Chairman] and [the Virginia Water and Thorpe Branch Chairman] ... and [the Surrey CF Chairman] would also like to meet to discuss the CF plan... (Ridge-Newman 2008a).

After the meeting took place on 23 September, I sent to the same recipients an email stating:

> I wish to thank [the Surrey Conservatives Chairman], [the Surrey CF Chairman], [the RWCA Chairman], [the Virginia Water and Thorpe Chairman) and [the newly appointed Runnymede and Spelthorne CF Deputy Chairman] for their attendance at and contribution to yesterday's meeting in Runnymede. I am sure we all agree it was a worthwhile meeting and a productive one at that. In summary, we have formed a new Conservative Future branch of which I am chairman. "Runnymede and Spelthorne Conservative Future" (RSCF) [later becoming Runnymede and Weybridge Conservative Future (RWCF)] will serve young Conservatives with strong links to the area. This branch will run independently of the collegiate branch of Royal Holloway, University of London. Although, we hope, the two will run in a complementary capacity. We are planning a membership drive over the coming months which will culminate in a drinks reception for prospective members. This is likely to be held on the evening of Thursday 22nd January, 2009, at the Runnymede Hotel with a prominent Tory politician as speaker ... We intend that invitations will be sent out two weeks prior to the event. The event will be free to attend... This event has been modelled on the recent success of the Elmbridge CF membership drive in which 29 new members were recruited... I will soon produce a Facebook group for members to join (Ridge-Newman 2008b).

As Smith notes, organizing something along these lines using the internet in Dumfries, in the run-up to the 2003 Scottish Parliament elections, would have been challenging, because 'the communications infrastructure that was needed to support broadband and internet connections was very poor' (2011: 53). In fact, Smith goes on to claim that broadband was a generally absent medium in much of the area at the time.

In the Surrey case, the discourses and interactions that surrounded events noticeably differed in terms of the preferred methods of communications used by the individuals involved. My dialogues with the younger members occurred almost exclusively using some form of internet application like email or Facebook. However, the older members preferred to be either called via telephone or to have a face-to-face meeting. This indicates that there was some division in the communication practices

DOI: 10.1057/9781137436511.0009

between the younger and older members. There is further evidence for this in the response that the RWCA chairman gave in an interview:

> The internet means different things to different people. There is an age divide here and there is a digital divide. The internet started off as a means of sending email messages, and to the younger generation now it's an enormously power-ful communication device using new media tools, such as blogs, Facebook, Twitter and so forth. The difficulty that we have in the Conservative associa-tions is that they tend to be populated by two sorts of people. There are the young and enthusiastic political volunteers, who join in their twenties, they work hard and they pass through. We then have a cadre of people who are there always. These are people in their fifties and sixties and seventies. The difficulty that many of these people face is that they may struggle when it comes to putting a new cartridge in their laser printer. So, they're perhaps not entirely hands-on. The internet, to an association that is run by this older cadre, is probably going to make less effective use of what the internet offers than might be the case, for example, in a young business or with some of the younger volunteers (Respondent Three).

Generally, this would suggest that the use of internet in the wider membership of the RWCA was, at that time, relatively generation specific. Similarly, in his ethnography, Smith (2011) notes that 'elderly' volunteers were less willing to canvass urban areas. The 'cadre' of older members in Runnymede, who are suggested to have been the control-ling body of the association in the run-up to GE2010, had an already established tradition and culture in terms of their uses of communica-tion technologies. Therefore, the use of new internet media by younger members was a divergence from the already established organizational culture of their associations. This can be explained by Schein (2010). His theory suggests that the culture of organizations in the postmodern or globalized contexts is being impacted on by the encroachment of what he calls 'contemporary networks', which result in a divergence from the more unified models of 'traditional' hierarchies that were more charac-teristic of the organizational culture observable in late modernity.

However, in the case of RWCA, there is evidence to suggest that, although the cadre of the party were less inclined to adapt their behav-iours and practices in line with technological advances, the roles of organizational elites, like the chairman himself, benefitted when useful internet technologies were embraced.

> As far as Cherry Orchard is concerned, which is our association office in Staines, I do work on the Cherry Orchard computers, probably one to two

DOI: 10.1057/9781137436511.0009

hours every day, but I probably only visit Cherry Orchard, apart from meetings, about four or five times a year, so almost all my work is done remotely and I suspect that this will be the model going forward which will have a series of agents who travel around the patch hot-desking from where they go. The internet has been such an enormously powerful force over the last ten years and what fast broadband links, or relatively fast broadband links, both where I live in Englefield Green and in Staines, mean is that I can sit at my desk in Englefield Green, I can use something like 'Remote Desktop' or 'LogMeIn' to log straight on to the Cherry Orchard computers. I can print material out on the Cherry Orchard printers or I can print it out at home. I just don't have to travel every day down to Cherry Orchard (Respondent Three).

This suggests that the role of internet technology in the daily working life of an association chairman is dependent upon personal choice; technical knowledge; access to the relevant technologies; and inclination towards the use of technology. These leadership qualities and choices are likely to impact on the manner in which the association and local campaigns are organized.

In comparison with my observations of the Anglesey Conservatives Chairman, the approaches of Conservative chairmen can vary significantly. For example, the Anglesey chairman, a busy local farmer, did not use the internet whatsoever. He delegated all administrative activities to association officers and answered a mobile telephone only on rare occasions. Furthermore, Respondent Three's testimony suggests that, during his chairmanship, his grasp for technological knowhow enhanced and changed his personal working experience within the party in a practical way, which is again a return to signifying the importance of the role of individual leadership (chapters 2, 3 and 4). The use of internet applications had reduced considerably the frequency of this chairman's journeys to the association office. Stuart Ball (1994a) writes that historically in Conservative associations the most important role has been that of the association chairman, because they are the driver behind the effective daily running of the local party organization. A lack of current scholarly perspectives on contemporary local associations means that this traditional view of local leadership is possibly dated (Whitelely et al. 2002). However, the RWCA case does suggest that there are, at least, characteristic remnants of the traditional role of the chairman extant in present day associations, which can be the deciding factor for the impact of the uses of new technologies, and therefore the effective embracement of contemporary networks, in some local settings.

DOI: 10.1057/9781137436511.0009

The use of internet by the RWCA for campaigning in the run-up to the 2010 General Election remained limited in the early stages of development as a communication tool for reaching the voter.

> We made very little use of email. We only have around 1,000 email addresses. We did try to work up some quite thoughtful email letters but it was early days for us. It's a full scale endeavour communicating in this way, as it's also a full scale endeavour in maintaining a web presence. It's easy enough to create a web presence but you then have to maintain it and that requires continuing effort, which you are not necessarily going to get from a voluntary party (Respondent Three).

The applications of internet technologies vary significantly in terms of the level of specialist knowledge needed to integrate them into the needs of an organization like a Conservative association. The RWCA chairman is indicating that, unless an association could have afforded to pay an internet specialist or had a willing volunteer with the technological knowhow and creative abilities to maintain the local party website, the use of a website at the association level for campaigning becomes significantly limited. Furthermore, his testimony suggests that, in order to have made direct contact with voters in the Runnymede and Weybridge Constituency via email, the association would have had to have taken a more innovative approach in order to have harvested email addresses from electors.

Digital barriers and digital freedoms

There were limits to the services, like website initiatives, provided to associations by the central party in the run-up to 2010. Respondent Three claims that:

> The party has quite an effective central website and they ... put a lot of energy into this over the [2009–11 period]. They have improved it considerably. Central Office [CCHQ] made a content management website product available to the associations. It was pretty clunky, it looked pretty old fashioned, it was just about adequate for putting up factual material. I think that the Conservative product that we had last year was a very old fashioned, passive, fact based website. It's going to be difficult for associations because everyone enjoys setting up websites, but they do not enjoy maintaining them so much. I think the energy for this will come from Central Office.

My observations of RWCA noted how the RWCA chairman made at least four attempts to secure voluntary assistance of younger members for the running of the website at Cherry Orchard; and that turnover of the volunteers was high due to individuals being unable to commit the time required to the frequent demands of the role. This and Respondent Three's testimony suggests that unless an association had an able and committed webmaster within the local organization, it would have had to invest significant funds into website development, and ongoing maintenance, or make use of the limited facilities provided by the central party in the run-up to 2010. Moreover, the digital and age divides contributed to attitudinal barriers within the local party organization that were likely to have slowed the developmental progress of RWCA's online presence. Research by Whiteley et al. (2002) indicates that 'age' and 'attitudes' of Conservative Party members have been longstanding variables that impact on the nature of local associations.

Respondent Three's testimony suggests that campaigning via websites at the local level was not a significant concern for investment by the central party in GE2010 terms. Although CCHQ had invested in a sophisticated parent website at Conservatives.com, the aspiration to have a contemporary web-presence at the association level had to be initiated and developed by the associations themselves. However, in the case of the RWCA, the likelihood of that occurring was low because of its traditional and ageing cadre. Furthermore, with Runnymede and Weybridge being seen as a Conservative 'safe seat' in GE2010, investing in a website was not as significant a priority as it was for other Conservative associations and their PPCs in more 'marginal seats', like, for example, Laura Sandys who contested and won the South Thanet constituency from Labour; and Robin Walker, who did the same in Worcester. Unlike safe seats and less winnable seats, marginal seat candidates were likely to be the recipients of Lord Ashcroft's £3 million marginal seat fund. This enabled Conservative PPCs in marginal seats the opportunity to invest greater resources in a more sophisticated campaign.

When asked about the role of the internet in party organization, Respondent Three claimed that the association 'used it extensively'. This is in contrast to his response on the use of the internet in terms of reaching the voter in GE2010. Respondent Three describes how internet technologies impacted in changing the association's campaign procedures. Firstly, he outlines the association's approach used prior to the GE2010 campaign:

> During the course of an election, the branches will collect canvass information that has to somehow get onto the central MERLIN system. Fresh canvass cards have to be handed back to the branches and then, on Election Day, this

DOI: 10.1057/9781137436511.0009

information has to be married up with the telling results coming from the polling stations, so that in the evenings we can identify the people who are our supporters and who have not voted. This is a classic 'get-out the vote' campaigning technique. In the past, this has all been paper-based. The branch would fill in the canvass card, it would then motor the canvass card over to the association office and this was the reason why you wanted an association office that was relatively nearby and relatively accessible. Volunteers would then put this into the computers and that information would then go out to branches.

The chairman goes on to describe how this was different in the GE2010 case and how the association utilized the internet innovatively in their campaign organization:

> We built an internet application that sat on top of Blue Chip, so that our branches were able to enter in their canvass information directly from their branch office, which was usually a bedroom or an office in somebody's home. They were able to pick up the results immediately, they could enter the telling results in immediately and they could then see who the people were who needed to be knocked up. Because this was all networked into the servers at Cherry Orchard, the telephone team at Cherry Orchard was able to see the same information in real time. We were able then to direct the telephone team at Cherry Orchard to back up the key stress points in our constituency and in the neighbouring constituencies that we were supporting.

This demonstrates the heterogeneity of Conservative associations, which has been attributed traditionally to their autonomy (Ball 1994a). Furthermore, this testimony shows how individuals can impact on the direction and culture of each campaign. The system described above was unique to RWCA because it invested the highly developed computer programming skills of the chairman. In doing so, the chairman used the internet as a platform to assist in innovations over and above that which had been supplied to the association in the form of the centralized MERLIN database.

Unlike MERLIN, its predecessor 'Blue Chip' was not designed with the potential for internet networking capabilities. Both databases were designed primarily for managing canvass information. In turn, this data is used to 'get-out the vote' on election days through a process called 'knocking-up', whereby Conservative activist from local associations and branches target declared Conservative voters to ensure they have used their vote at their local polling station. Smith (2011) notes that in Dumfries, c. 2003, there was access to only one Blue Chip computer,

DOI: 10.1057/9781137436511.0009

which regularly 'crashed' while helpers were inputting the canvass information. MERLIN was different to its predecessor in that it allowed the potential for the data to be used also centrally through a national MERLIN network which was fed directly via the internet to a central database at CCHQ (Chapter 2).

The extent to which the advent of political internet technologies had shifted the canvassing culture of RWCA is illustrated in the following narrative:

> The traditional approach from about 1997, when Blue Chip was first issued to associations, was that each branch would have its own little committee room on campaign day. There would be one PC operating there, the telling results would be coming in to the committee room and then that computer in the committee room would be printing out lists for people to go out and actually bang on doors to get out their vote. In some ways it is the same but in the pre-internet days every branch committee room was its own little silo of information. In the 2010 context everybody's connected, which means there's mutual sharing of support and support can be directed to the point at which it is most needed (Respondent Three).

This shows how for RWCA the internet loosened to some extent the otherwise isolated approach to the sharing of campaign information. However, the RWCA chairman's experience in 2010 was in stark contrast to the organization and campaign techniques I witnessed in the Anglesey association as their candidate on election day. Anglesey Conservatives did not have the political will, resources, volunteers or organization to conduct even the simplest telling/get-out the vote operations and, therefore, there was comparatively very little use of internet technologies by Anglesey Conservatives in election day procedures. The comparison of the Anglesey and RWCA cases demonstrates further the extent of the heterogeneous nature of each Conservative association.

An example of how individual associations can be different in their use of internet technologies is demonstrated by the approach of the Spelthorne association, which constitutes the other half of the RWSCG. Unlike RWCA, the chairman of the Spelthorne Conservatives sent out 40 'eNewsletters' between 3 November 2008 and 5 May 2010. My analysis of the contents of these emails suggests that their intended purpose was to inform the party supporters listed in the association's email database of news, information and events that pertained to the elections and organization of Spelthorne Conservatives. The emails were sent also to some non paid-up members. I know this because I continued to receive these emails

DOI: 10.1057/9781137436511.0009

long after my membership with the Spelthorne association had lapsed. In the run-up to GE2010, I received eNewsletter-style emails sporadically from Kingston and Surbiton Conservatives; regularly from PPC Shaun Bailey and Hammersmith Conservatives; regularly from Cities of London and Westminster Conservatives; and a prolific range of emails from PPC Zac Goldsmith and Richmond Park Conservatives. Goldsmith's campaign was the local Conservative target seat that was assigned to the RWCA in a relationship which the party calls 'mutual aid'.

Another association of which I had been a member was Cities of London and Westminster. I had joined as a CF member, paying £10, at a CF event in London. Subsequently, I began receiving email communications from the association about Conservative social events. I continued to receive communications of this nature long after my membership lapsed. However, in November 2010, all email communications from the association ceased. Richmond Park collected my email address on a number of occasions, including at their social events and on the days that I campaigned for their candidate. Subsequently, I received duplicate emails from them, which as a recipient became an irritant. The other associations mentioned above added my email address to their email list after I emailed them to offer my time to their campaigning activities. All these associations, except Spelthorne and Richmond Park, ceased all communications with me soon after GE2010.

These cases collectively demonstrate significant heterogeneity and autonomy in the manner in which the observed Conservative associations collected email addresses and disseminated internal news via email. This would suggest that Cameron's Conservatives in the associations were free, to some extent, to decide upon their own style of email strategies and use. Furthermore, individuals in leading positions within the association played a key role in how email was used and disseminated. Based on these observed cases, the frequency of emails sent would appear to have had some correlation with the desires of the candidate, the association, and the wider party for that particular seat, to win at the general election. Richmond Park, as a key target seat for the Conservatives, seemed more aggressive in its approach to email distribution than the nearby safe seats of Runnymede and Weybridge, and Spelthorne. Richmond Park Conservatives rallied extensively its supporters through the use of email and Facebook. Interestingly, when compared with Kingston and Surbiton Conservatives, and Hammersmith Conservatives, whose approach to using email to

DOI: 10.1057/9781137436511.0009

organize their supporters was milder, Richmond Park had the greatest electoral success in GE 2010 (Dods 2010: 179, 271, 615).

Where the target seat of Richmond Park used innovative techniques in internet campaigning through active methods for collecting email data, the safe seat of Runnymede and Weybridge campaigned in a more traditional way and admitted that it did not make a significant attempt to use email. Therefore, it would appear that at the association level, the likelihood of innovative use of internet applications was dependent on the technological ability and/or inclinations of leading figures within the association, for example the candidate or the chairman, and the status of the seat, for example whether or not the seat was a Conservative target seat. However, both associations significantly used internet applications to organize their supporters.

Virginia Water by-election campaign: 4 June 2009

In the RWCA, campaigning in the run-up to the 2009 European and local elections began in March 2009. In early 2009, I had entered into an informal dialogue with the RWCA Chairman and the Virginia Water and Thorpe Branch Chairman about the possibility of my standing as a Runnymede Borough Council candidate. There was an opening for a new Conservative candidate in the Virginia Water Ward, because an incumbent councillor in the Ward had intended to resign. The candidate's resignation was managed and timed by the association cadre in order that it triggered a by-election to coincide with the dates of the other local elections scheduled for 4 June 2009. My eligibility and suitability to fill the role was discussed by the more senior cadre of the association.

The process of candidate selection began with a letter of notification for a face-to-face interview, which was sent in the post. The interview panel was comprised of senior association officers and representatives who were assembled by the association chairman. Before I could be fully approved by the Conservative association, I was instructed by an email from the association chairman, sent 7 March 2009, to arrange a meeting with the Conservative leader of Runnymede Borough Council. Through this process, I became aware that there were two separate but equally important hierarchical structures at work in Runnymede in terms of party organization. I noted above the significance and influence that leadership figures can have at the local level. In Runnymede, there

DOI: 10.1057/9781137436511.0009

were noticeable factions of Conservative participants who had subtle inclinations to side either with the association chairman or the leader of Runnymede Council. According to Ball (1994a), Conservative associations have long held the potential to form 'cliques and factions' that can lead to manoeuvring and sometimes intraparty battles. This phenomenon was accentuated in Runnymede because of the large number of Conservative members on the Borough Council being Conservatives (86 per cent on 4 June 2009), thus giving greater weight to the political arm of the local party.

Once approved by the Conservative leader of the council, the branch members overseeing the safe seat of the Virginia Water Ward agreed to my candidature in consultation with the branch chairman. In the process of becoming a local candidate in Runnymede, the internet, in the form of occasional email exchanges between the candidate and local Conservative leaders, played only a minor role. Email was used as a tool for communication, which, in turn, facilitated minor organizational aspects of the process, like introductions, meetings and interviews. However, this was no revolution for the Conservative participants involved in the process. Email attributed little more to the process than what would have been achieved using a telephone call or text message. Therefore, the selection of a local candidate remained relatively traditional in its approach.

The selection was a stepwise process with checks and balances in place that distributed the power for the adoption of the candidate between a number of key Conservative leaders and the collective approval of their respective cadres. These cadres together collectively formed a wider local cadre with factions forming around the key leaders in the local organization. Email as an additional tool for communication oiled the processes of interaction between these individuals and groups in the process of candidate selection in that it assisted the flow of the internal procedural mechanisms. However, where local candidate selection remains within the hierarchical control of the local political and organizational elites, the scope for the use of internet technologies in that process is likely to continue to be limited.

Being a prospective local candidate, I became elevated in the hierarchical ranks of the local Conservative Party and, subsequently, found that the active members of the local party conveyed a greater sense of trust towards me. As research by Dirks and Ferrin (2001) suggests, 'trust' plays a significant and beneficial role in organizations. However, in the Conservative Party case, this is balanced by its hierarchical tendencies

DOI: 10.1057/9781137436511.0009

in which one's behaviour is tempered by a sense of watching-eyes in the higher ranks of the party. A latent culture of deference to those in higher authority is a traditional characteristic of the party (Ball 1994b). This phenomenon remains inherently cultural to the Conservative Party and is one that I observed throughout the ethnographic experience at the local, devolved and national levels. It is a principle which is passed-on to newly initiated members. Often these members are eager to please the party collective and, therefore, subsequently fall-in-line. There is an unspoken but tangible understanding among party office-holders of order and compliance which emanates through the day-to-day actions and interactions observed within the party collective. These phenomena had repercussions in both the on- and off- line world.

Digital organization

At the association level in the run-up to 2009, I observed and experienced a significant distrust for the use of email to engage with unknown entities/individuals on matters that were potentially sensitive in nature. However, as my time within the party accumulated, I journeyed closer towards becoming a figure within the association cadre who was accepted as someone to be 'copied-in' on certain emails in relation to the organization of the party and its campaign strategies. It was an unspoken and informal process which evolved over time. It appeared to be led by the local leaders like the association chairman and the leader of the council. When a local party leader signalled in an email to a staff member, officer, councillor or activist, through the action of copying me into the email, or introducing me via email, I found that my email interaction with party participants was exchanged more freely and less guarded in its content.

The prominence of a traditional cadre within RWCA means that new individuals have to learn the cultural ways of the association; and invest time and other resources in order to, firstly, become initiated and, secondly, develop trust relationships. It is the leadership of specific individuals within the cadre who can use the simple gesture of openly including a new Conservative participant on the email distribution list of an internal email. Once that action is commenced by the individual in the leadership role, others in the cadre receive this as an affirming signal to follow suit by cautiously accepting email interaction with the new individual. However, that point is not likely to be reached unless the new participant has regularly interacted with key individuals within the association or political cadres at local Conservative Party social and

DOI: 10.1057/9781137436511.0009

political events. Therefore, an introductory email, or the appearance of a newly trusted name on a distribution list, acts as an electronic symbol of acceptance and trust, which is likely to have been earned first in the offline world. Whitty and Joinson (2009) state that the internet is often perceived as an 'untrustworthy space'. As Tanis and Postmes (2007) suggest, this results in face-to-face interaction being 'superior' in comparison to less personalized mediation of the internet. Based on my observations, these factors were integral to the customary behaviours, of the older cohorts in RWCA, who were the orchestrators of the processes for initiation through which newer members learnt the ways of the association.

I found that 'trust', to some degree, was extended to my campaign team – as though they were an extension of me. As I had contributed significantly to the founding of RSCF, I had a number of young Conservatives from the branch who formed my Virginia Water campaign team. Therefore, because these party workers were an extension of my contribution to the party, they became relatively trusted and embraced by the chairman of RWCA. The trust and rapport I had worked to develop was inherited by my team members on an individual basis. For example, on 13 March 2009, my campaign manager, an 18-year-old Conservative, found himself being copied-in on emails from the association chairman on matters of telephone canvassing at Cherry Orchard. The chairman of the association entrusted the younger member with the responsibility of managing the telephone canvass and recruiting other local CFers for the activity over the medium of Facebook. In this case, the internet was used in a liberal manner by the chairman because a relationship of trust, albeit secondary, had been established already between him and the candidate. Therefore, the candidate's selection acted as a validation of the candidate's judgement in others. Those individuals that the candidate brought into the Conservative-fold, as campaign team members, were automatically given a degree of trust to manage campaign matters using internet technologies by the virtue of the candidate's discernment.

While Facebook was used regularly by the younger members of my campaign team to organize individuals and groups of CFers for activism in campaign events, like door-to-door and telephone canvasses, the older members of the Virginia Water campaign team used email. As a relatively new member of the RWCA, I was meeting members and councillors on a continual basis. Some of these introductions were made in person and others using email. On 1 April 2009, a fellow candidate introduced himself to me via email. My email address had been provided to him by a senior official in the local party. This was a relatively common practice among the established and trusted local elites in the association. Therefore, in this manner, email was acting as a medium that facilitated more fluid network interactions between local elites, as they rose to positions of relative seniority within the association. Moreover, this is further evidence for the digital segregation or 'divide' between the younger and older Conservative participants, to which Respondent Three and others have testified.

Trends in the choices and uses of internet technologies in the RWCA appear to have been dependent on the two distinct age groups in the association. In terms of activism, there was a small but keen group of younger Conservatives who leant towards the use of Facebook to organize their involvement within the local party. However, they also used regularly email and mobile telephones. Often, their mobile phones were used as devices with which internet connections were made in order to access email and Facebook. That said, these were the days before the widespread use of smartphones, therefore much of the mobile cellular technology was of more primitive forms of interface. Some young Conservatives used their phones to update their Twitter status with short microblogs, within 140 characters, detailing their activities on the campaign and other political messages.

The CFers created a Facebook group for paid-up members of RSCF, which was also extended to unofficial supporters from outside the local area who could request to join the closed Facebook group. The group was typical of those administered by other CF branches, which utilized Facebook groups in order to grow a network of supportive and active Conservative participants; share organizational, campaign and other political information on the group's Facebook Wall. Often, this would encourage debate through online activity involving group members interacting with one another and the medium by posting comments in public conversational threads. Facebook was used to organize also campaign events, social events and other organizational and political activities. Rachel Gibson (2013) suggests that this type of activity presents a new model of grassroots campaigning that has led to the devolution of power from the professional centres of party organization, because of the low financial cost of engagement in social media.

The freedom from financial burden in using Facebook as a marketing tool certainly enhanced the Virginia Water campaign. The administrators of the Facebook group were able to send Facebook messages to the entire group, which acted as an instantaneous and targeted promotion device during the campaign period. The chairman of RSCF regularly sent messages to the members of the Facebook group in order to encourage them to attend specific campaign days. Individuals were invited to attend political, social or campaign events through the creation of Facebook event pages. The event page would give information about the time, date and location of the event in addition to further information and a list of those who intended to attend the event. The members of the RSCF Facebook group were subsequently digitally invited to the campaign event using simple Facebook functions. Members of the group were encouraged to 'RSVP', thus showing whether they were attending/not attending/maybe attending. On 9 May 2009, 12 members of CF, including local members and members from outside the local area, came to a Virginia Water 'action day'. Most of these individuals had interacted with the group and the event information on Facebook.

In addition to the CFers, there were other, more senior, activists who attended the action day in Virginia Water, however, they were fewer in number and their participation in the campaign day was organized via other media, namely email and telephone. None of the senior activists were members of the Facebook group. Therefore, they did not receive notification of the campaign day via any Facebook interaction. It is not possible to know who, and to what extent, the senior members had interacted already with Facebook in their personal lives, but as an administrator and observer of the Facebook group, I was aware of who the group members were. Both chairmen of the RWSCG joined the RSCF Facebook Group in the run-up to 2010, and sometime after the Virginia Water campaign.

During the Virginia Water Action Day, the RWCA Chairman and another senior local Conservatives expressed how impressed they were with the turnout of younger people and acknowledged the role that Facebook had played in the successful organization of the event. However, one senior local Conservative, who assisted in the organization of the canvassing part of the action day, made disparaging remarks about the young people who 'turned-out' to help in his 'patch'. His concerns were centred on the lack of control he had over the organization of the campaign day, which was a new and uncomfortable change to the manner in which the branch had executed its campaigns in the past. In a campaign day debrief on 11 May 2009, the same Conservative, who at that time was a septuagenarian, said that the best way to contact him was by telephone and that he was not of the generation who felt the need to communicate regularly by email. This account suggests that the digital divide in the culture of the RWCA could cause, at times, a clash between the generations. Perhaps the digital divide in RWCA might have been narrowed if there had been a wider understanding of the benefits of internet use in political organization at that time.

The political networking capabilities of Facebook were demonstrated when a South East Region European Election candidate searched for my name on Facebook, found my personal profile, and sent me a Facebook message offering to help me canvass in the Virginia Water Ward. At that time, I had not had any previous contact with the candidate whatsoever. Therefore, through having a presence on Facebook, as a local election candidate, I made myself and my campaign accessible for assistance from Conservative participants outside of my local area. Furthermore, through having a basic but informative campaign website which included my email address, I received an email from an assistant to the Conservative South East MEP Nirj Deva who offered me access to his campaign team. Subsequently, Deva and his large team of young supporters arrived in Virginia Water to assist with door-to-door canvassing on what they called Nirj Deva's 'Battlebus'. These examples demonstrate how, in 2009, the online presence of a relatively unknown council candidate could be converted into significantly tangible offline interactions through the power of the internet. In this case, the internet acted to significantly enhance Conservative campaign activity in Virginia Water, while investing only minimal financial resources in the technology.

DOI: 10.1057/9781137436511.0009

7

Anglesey Conservatives and the Internet

Abstract: *Chapter 7 features the run-up to GE2010 through the eyes of the researcher as the PPC for Ynys Môn. It is argued that although the internet was used to enhance the campaign in some circumstances, for instance, the use of the medium to facilitate a remote candidate presence, and a virtual campaign team of participants separated by large geographical distances, the internet, in particular e-campaigning, was not a priority for use within the campaign. In fact, although the internet helped with organization and process, the lack of sufficient internet capacity in some aspects of the party's wider national campaign organization led to negative bureaucratic impacts at the local campaign level for the candidate.*

Keywords: Albert Owen; Anglesey Conservatives; Anglesey | Ynys Môn; Plaid Cymru; Welsh Conservatives; Welsh politics

Ridge-Newman, Anthony. *Cameron's Conservatives and the Internet: Change, Culture and Cyber Toryism.* Basingstoke: Palgrave Macmillan, 2014. DOI: 10.1057/9781137436511.0010.

DOI: 10.1057/9781137436511.0010

Shortly after the Virginia Water by-election, in summer 2009, I applied to the Conservative Party to become a member of the approved list of parliamentary candidates. I was invited to attend a Parliamentary Assessment Board ('the PAB') and, in the August, I was told that I had been approved to become a member of 'the Candidates List'. I applied to a number of Conservative associations that had a vacancy for a PPC in their constituency. In early January 2010, I received an email from Ynys Môn | Anglesey Conservative Association inviting me to a selection meeting on 9 January 2010. At the meeting, I was elected by the association's 'paid-up' party members as their candidate to represent the Conservative Party in the impending General Election, 6 May 2010. The following month, I moved to the constituency fulltime and resided in Trearddur Bay, near Holyhead. There, I coordinated my parliamentary campaign and ethnographic fieldwork in a simultaneous and complementary capacity. This chapter details that experience and analyses the role of a range of communication technologies in the local campaign.

Ynys Môn parliamentary election campaign: 6 May 2010

Some argue that British communities are more alike than we would generally suppose (Edwards 2000; Smith 2011). Others suggest that heterogeneity is a more prevalent feature (Cohen 1982). However, it would seem that both perspectives can hold truth within the same case study and are dependent on the standpoint from which one is viewing the subject. It is the reason why this book has favoured complex descriptions rather than simplistic parsimonious models to investigate phenomena. Furthermore, these factors are in themselves variables that influence the manner in which the case of interest is explained. For me, as both an ethnographer and politician, Anglesey was culturally different to the life I had experienced in Surrey. Anglesey being an island meant that it had a distinct identity, culture and community of its own. I had to considerably adapt my approach to people and politics in order to be accepted within, what was to me, a new culture. I spent some time as both a politician and ethnographer observing and learning the culture, customs and everyday lifestyle of the islanders and the Conservative association. I began learning the local language of Welsh and adopting the ways of the 'Anglesonian'.

DOI: 10.1057/9781137436511.0010

During that process, I developed a stronger sense of what it meant to be a Conservative PPC. It was clear to all concerned that I was not a 'true local'. However, I found that, in general, the local people of Anglesey accepted my candidature in a welcoming and hospitable manner. Many members of the Anglesey association gave me a very warm welcome. After my selection, I was immediately invited by some of the association members to a country cottage for 'bacon butties'. My first impressions of Anglesey life was its remoteness and its areas of relative poverty in comparison with Surrey. Some islanders prided themselves on Anglesey being undeveloped, while others blamed local and national politicians for a lack of opportunity on the island. Anglesey's simplistic infrastructure led many locals to claim that the island was an aesthetic and cultural remnant of 1950s Britain. However, I felt that the islanders' innate perception of being an isolated community may have fuelled the sense that they were more 'hard done by' than other parts of Britain. So while the island did have some things that made it distinctive, it was not perhaps in certain ways so unlike other comparable parts of Britain like, for example, Cornwall, Northumbria, Skye and Thanet.

In 2010, one cultural symbol of contemporary life which had been noticeably slow to develop on Anglesey was high speed internet, with the island's high number of 'not-spots' and 'slow-spots' (IACC 2012). As the Conservative candidate, I received a number of complaints from constituents and local entrepreneurs who perceived Anglesey to be a place that had been neglected in terms of the development of the island's internet infrastructure. The poor access to broadband was considered to be hindering the development of businesses on the island (Williams 2010a). Rather than using email, the relatively aging population of the island, including the local Conservative association which had no more than five members under the age of 40, preferred to communicate by telephone and, occasionally, by letter. Therefore, in terms of my parliamentary campaign, the use of internet technologies to reach the Anglesey electorate was, relatively, a low priority. Placing this in the wider context, Southern and Ward (2011) show this finding is in keeping with the general trends in e-campaigning observed across the UK in GE2010. Moreover, it would appear that there was no significant correlation between candidates' use of e-campaigning and electoral success in GE2010 (Williamson 2010).

I focused much of my attention on devising and delivering traditional door-to-door leaflets; writing letters to the local newspapers; and

DOI: 10.1057/9781137436511.0010

achieving media and press attention like my appearance on the BBC *Politics Show* and interview with BBC Radio Wales. This approach is not dissimilar to the account given by Alexander Smith (2011), which details the campaign approach taken by the Scottish Conservatives in 2003. This would suggest that, although the use of internet technologies had significantly proliferated and begun impacting on campaign organization, grassroots Tory political culture, in terms of the pursuit of reaching the voter, had not significantly changed in the run-up to GE2010. Certainly not with any revolutionary impact. I spent the largest portion of my time canvassing door-to-door with members of my small, but dedicated, campaign team with the strategic aim of meeting as many people face-to-face as possible. This aim grew out of what was considered to be, by Anglesey Conservatives, an historical Conservative tradition on the island, which had been passed down in Conservative folklore since the time of the Conservative MP Keith Best. Keith Best was the most electorally successful Conservative candidate on Anglesey. Therefore, I made the strategic choice to emulate his mythological campaigning style.

I advertised a weekly surgery held at the association office in Llangefni, the island's county town, and made attempts to attract attention using local newspapers. I submitted press releases on stories about campaigns that I had organized in the various towns around the island, which included Amlwch, Beaumaris, Benllech, Holyhead, Llanfairpwllgwyngyll, Llangefni, Menai Bridge, Newborough, Rhosneigr and Valley. We placed half page adverts in the local newspaper. The adverts invited constituents to attend public meetings that were held in eight of the island's main towns. Jon Lawrence (2009) claims that attendance at election meetings has been in general decline since the mid-1950s. Anglesey Conservatives were aware of this trend and it came as no surprise that the numbers at each meeting were between 5–28 individuals. Interestingly, however, informants in the Anglesey association reported that attendance in 2010 was higher than the previous two general elections.

In the three weeks prior to polling day, I attended a number of official hustings, most of which drew audiences in excess of 100 people. These included Mencap Cymru; the Federation of Small Business; the Farmers Union of Wales and the National Farmers Union; and Churches Together. I organized a campaign weekend with over 25 participants who travelled from across the UK to join an 'Around Island Rally'. We spent four days touring the island using a 'loudhailer' to expound the Conservative

DOI: 10.1057/9781137436511.0010

message that it was 'Time for Change on Ynys Môn'. The loudhailer was especially useful in the small villages that we had not been able to canvass. The teams of supporters arrived in each village and posted through the door of each home a letter from the Conservative candidate and a short leaflet that detailed both local and national issues. As soon as the Labour Prime Minister, Gordon Brown, called the election, on 6 April 2010, a team of party workers spent a week erecting Conservative 'Ridge-Newman' posters/boards on private property across Anglesey.

Although the majority of my time was given to traditional campaigning, a significant amount of my day as a fulltime candidate was spent using computers and the internet from either my home or the association office. In the early days, soon after my selection as a candidate, I spent between 10 and 18 hours per day in front of a computer planning, organizing, managing, communicating, writing and designing aspects of my work as a candidate. Each candidate can have a unique experience, because of the dynamic variables which influence each parliamentary campaign. These might include: the candidate's style, skills, leadership abilities, experience, autonomy and creativity; access to political staff, funds and campaign support; and private and personal influences, like family support, external stressors, accommodation concerns and personal wealth. Other variables might include: geographical factors; travel concerns; temporal factors; local cultures; interpersonal dynamics between local activists, association officers, Conservative Party professionals and the media; and the mood of the electorate.

A number of fellow candidates reported in casual conversation at conferences how they felt that they had to be on their computer a lot. However, many of them did not value it as an activity that helped them win any more votes, thus favouring being 'out-and-about' meeting people in the community and on the doorstep. Butler and Kavanagh (1992) questioned the role of the local campaign in national electoral outcomes and argued that there was a lack of evidence to suggest that computer-based constituency campaigns had any significant benefit in 1987. In 2010, it would seem this could be argued to have been the case in the Anglesey association. Whereas, in the RWCA case (Chapter 6), the entrepreneurial approach and digital expertise of the association chairman acted in a manner that meant a computerized system did impact significantly on the 'get-out-the-vote' operations of their local campaign. Therefore, technocultural behaviour at the Tory grassroots seems to have been a mixed picture in GE2010.

DOI: 10.1057/9781137436511.0010

PPC for Ynys Môn and internet technologies

On becoming a Conservative PPC, I received a comprehensive 'Campaign Pack' (CCHQ Wales 2009) in the form of a large A4 ring binder with accompanying digital files stored on a CD-ROM. The pack was compiled and issued by CCHQ Wales and supplied to all candidates and agents in the region. A similar pack was distributed to candidates and agents in the other regions. The Welsh Conservatives campaign pack differed to the English pack because some candidates in Wales chose to incorporate Welsh language considerations in their campaign. For example, I stipulated that all literature in the Anglesey campaign would be bilingual. The packs predominantly gave advice and guidance to constituency campaign teams on traditional methods of campaigning, with the majority of the pack being focused on canvassing and traditional political literature. The pack did offer some advice on 'e-campaigning', which included an outline of the role of an e-campaign director within a campaign team; how to campaign online; and how to get-out the vote using email. Supplementary information on 'e-campaigning in an election' (CCHQ London 2010) was later provided by CCHQ London in the form of an election memorandum. The information was contained in a PDF attachment to an email. However, in contrast with the information in the campaign pack, the memorandum focused on the rules and legalities of e-campaigning rather than how to use e-campaigning techniques to win votes.

The Welsh Conservatives campaign pack states that: 'Online campaigning is the use of text messaging, emails, websites and social networking sites to communicate our message' (CCHQ Wales 2009). This is useful in that it encapsulates the Conservatives' 2010 corporate definition of e-campaigning. Furthermore, it would be telling about how the Conservatives perceive campaigning online by examining whether and how this definition is changed in subsequent elections. The party's rationale for using online techniques in 2010 was rooted in the notion that contemporary modes of communication are useful when used in conjunction with 'traditional campaigning methods' in order for the party message to have greater electoral reach (CCHQ Wales 2009). The guidance outlines also the benefits of using the internet. It states that email is cost effective and speedier at disseminating information. The pack encourages campaign teams to write emails with bold attention grabbing subjects and in a manner that expounds the central messages

DOI: 10.1057/9781137436511.0010

of the campaign. Emphasis is placed on websites as being 'essential' parts of contemporary campaigning so that local voters can gain access to information on the candidate and wider campaign activities. In line with the issues discussed in relation to WebCameron in Chapter 2, this demonstrates an increasing maturity in the central party's understanding of the benefits of using publically accessible communication channels that are entirely controlled within the party.

In terms of social media, the pack supports the use of social networks like Facebook and Twitter, stating that they can be a good way of informing those that do not read the party branded leaflet 'InTouch'. The pack suggests also that social networks can be a tool for finding new helpers. When compared to the party's approach to the use of Facebook in the 2008 Mayoral Election (Chapter 4), in which case the party's use of Facebook was evolving in an organic manner at the grassroots, it would seem that by 2010 the central party's understanding of social media was also beginning to mature in the sense that CCHQ felt able to promote its practice more confidently in the campaign context. The online section of the pack ended with a warning about the use of the internet and the pitfalls of publishing text online at speed – without the traditional process of 'sign-off' being in place. Instead, the online checks and balances had to be made by the individual through self-regulation and self-censorship, which placed both power and responsibility in the hands of party activists and candidates. This demonstrates that by late 2009 the central party was aware of both the pros and cons of online campaigning. However, the party took a relatively passive approach to managing the electronic output of campaign teams in comparison to the more structured and rigid approach of the sign-off process for traditional campaign literature.

As Fisher et al. (2011) write, the central party was particularly sensitive to erroneous 'spending commitments', and, although there was no sign-off process for web-based publication, the central party did monitor grassroots social media activity. I experienced this sensitivity first-hand, during the run-up to GE2010, when one senior Tory questioned a letter I had written to the then Secretary of State for Wales, which called for greater financial investment in Anglesey's local economy and had by-passed sign-off. This was viewed to be a spending pledge that had the potential to cause problems over the Dispatch Box. The Wales minister did reply to my letter, but did not raise the matter publically. Therefore, the case highlights the internal concerns, held by some in the upper

DOI: 10.1057/9781137436511.0010

echelons of the Conservative Party organization, about spending pledges and the anxiety that ensues when they feel that they are not in control of the grassroots campaign.

Internet and campaign geography

When selected as the Conservative PPC for Ynys Môn, I resided over 280 miles away in Surrey for one month before relocating to the island. I travelled twice to Anglesey on constituency business during my first month of being the PPC. Although these geographic limitations prevented me from having a daily presence in the constituency, in the early stage, the internet facilitated my role as a long distance candidate. During the first weeks as PPC, I worked up to 18 hours per day using internet technologies and applications via a laptop computer. I had 117 days between my selection and the election day, which was a greatly shorter period than many candidates. I was provided far fewer resources than target seat candidates in terms of funds, manpower and general support from CCHQ. Therefore, I invested heavily my own time, resources and skills to make up the shortfall. Initially, I focused on bringing together my political supporters to form a remote campaign team from all corners of the UK, many of whom were young Conservatives who had been displaced out of London and the South East because of either work or university commitments. Their geographic distances away from the island meant that they would not be able to travel to Anglesey regularly, but many agreed to take on campaign team duties from afar. I communicated with these individuals through Facebook and email, but also using SMS or 'text messages' on my mobile phone. I held two face-to-face meetings at the Carlton Club with a political communications expert. These meetings were organized using Facebook and SMS.

Using internet-based discussions with members of my remote campaign team, I was provided with technical and creative advice on aspects relating to social media and website design, in addition to other offline aspects of the campaign, like strategy and fundraising. Each member of the team was assigned tasks using textual instructions either via email or Facebook messaging. Web posters for my personal campaign website were developed with discursive exchanges over email in order to discuss amendments to the design. I personally designed and built a basic website, using a free website building application at Webs.

DOI: 10.1057/9781137436511.0010

com. Members of the remote campaign team contributed to the website in designing and submitting to me, via email, photographic based posters, banners and logos. These were designed using Adobe software and sent to me in image or PDF format, which I uploaded to the website.

The internet facilitated a virtual campaign office in which I was able to conduct campaign team meetings, relatively in real time, across geographical distances. Despite living a significant distance away from the constituency, the internet facilitated my role as a candidate, allowing me to comprehensively plan and organize my campaign strategy in advance of my move to the constituency. To some extent, my early role could be described as a virtual candidate insofar that I conducted my first month as the candidate remotely from Surrey. I used internet technologies to help me campaign, communicate and organize as if I were already resident in Anglesey. Facebook and email acted as both private and open meetings in which I could invite one individual or a number of individuals to engage in discourse in relation to specific campaign ideas and practical tasks. The internet-based processes were not substitutes entirely for face-to-face interaction, through which, as a candidate, I could have more effectively expressed emotions using expressions, gestures and social cues in relation to tasks, for example conveying a degree of enthusiasm and passion for the campaign (Mok et al. 2010).

Such interpersonal interaction would have had the potential to stimulate camaraderie between campaign team members, which was an ingredient to a successful campaign that lacked in the remote campaign team. Instead, members of the team worked in isolation, which, rather than building stronger interpersonal relationships between the involved individuals, led to a focus solely on a task completion mentality, which, although efficient, was not conducive to the sustainability of the remote campaign team. Therefore, in the run-up to my relocation to Anglesey, one-by-one the once committed members of the campaign team began to drift out of regular contact. This body of virtual campaign assistants was gradually replaced by a denser concentration of face-to-face interactions with a constituency centric campaign team constituted by members of the local Conservative association and supporters whom I drew in through my interactions within the constituency. These findings are in keeping with those of Whiteley et al. (2002) whose research emphasizes the importance of healthy local organization, even in the age of television, in improving the electoral chances of political parties. It, therefore, tempers the notion and validity of a cyber party (Margetts 2006) system in Britain; and suggests that although the internet is a valuable tool for enhancing aspects of the local campaign, it is not a suitable replacement for a strong local supporter base.

Bureaucracies, campaigning and the internet

The nature of the daily role of the Conservative PPC in GE2010, which involved a heavy reliance on bureaucratic exchange and communication via internet technologies, namely email, meant that the isolation described above was exacerbated by the volume of internet-based candidate responsibilities. A Conservative candidate in GE2010 who did not possess at least basic email skills would have been limited in their ability to fulfil the expectations that the central party at CCHQ London had placed on its candidates. Over my 117 days as PPC, the frequency of emails received in my campaign inbox, via my ant@politician.com, Mail.com, email address, increased significantly as time progressed. Between 6 March 2008 and 28 May 2010, I received in total 4425 emails in relation to my various roles within Cameron's Conservatives. The average frequency of emails received per day, inclusive of weekends, during that period was 5.44. In comparison, the emails received in direct relation to my parliamentary campaign in Anglesey, between 9 January 2010 and 25 May 2010, averaged a frequency of 20.81 per day. On busy days, my inbox received up to 90 emails within a 16 hour period. Many additional emails could be archived immediately. However, an average of 21 emails per day needed time and specific attention. Emails from the Conservative Research Department, the Welsh Conservative Press Office and the regional and national campaign headquarters often required intensive reading of attached PDFs. Respondent Eight testifies that, as a PPC, it was not appropriate to the campaign aims to give excessive time to reading all of the emails sent by CCHQ.

As the Anglesey PPC, approximately 36 paper-based letters were received from constituents compared to a significantly greater 186 emails. This was perhaps because of the relative cost effectiveness, ease and speed in sending an email when compared to traditional mail. However, one notable challenge for any candidate or MP is identifying the authenticity of the senders of email. In 2010, unlike traditional letters, it was not customary to include a verifiable handwritten signature or physical address in emails. Therefore, identifying whether the sender of an email

DOI: 10.1057/9781137436511.0010

is a legitimate resident of a politician's constituency was challenging. Some candidates and MPs chose to reply with instructions on automatic emails in the attempt to ensure an address was given. For example, the parliamentary email account of one Conservative MP sent automated receipts stating that any sent emails should include the sender's full postal address otherwise there would be no reply given. Some MPs chose to reply to emails by post in order to ensure that the address given was correct. This demonstrates how by 2010 internet communications had impacted in a way that some MPs had engaged in a process of rethinking their approach to corresponding with their constituents.

Second to emails, leaflet production was the most time intensive bureaucratic activity of the GE2010 campaign. By 2010, leaflet production relied almost exclusively on communication exchange via email. As a candidate without a professional agent, I was reliant on one part time administrator in the association office, a team of 16 part time volunteers, and occasional digital-based assistance from my remote campaign team. Therefore, in relation to my printed communications, I personally took much of the responsibility for photographs, text, organization, communication, liaison and printing. My leaflets consisted of an 'Introductory Leaflet'; 'Election Address'; generic 'Dear Elector' letter; and small glossy leaflet. I also advertised public meetings using a home printed leaflet and adverts in local newspapers. The process of leaflet production included: writing the text; having it translated into the Welsh language using a professional translation company; sending the English and translated text to the printers by email in Microsoft Word in order for them to insert it into a generic template provided by the Conservatives' literature pack; receiving the returned design proof from the printers; relaying back and forth between the printers in order to amend errors; emailing the final proof to a chain of Conservative officials, including the CCHQ directors in both Wales and London and the Shadow Welsh Secretary (for final sign-off); having the literature printed; having the prints delivered to the association office; driving by car the 40,000 printed leaflets to a Royal Mail Depot in England (not Wales); and, finally, having the leaflets delivered by Royal Mail to almost every domestic dwelling in the constituency. Therefore, leaflet delivery and production when multiplied by 650 Conservative candidates was a time intensive process for those candidates, campaign teams, CCHQ offices and Conservative officials involved.

DOI: 10.1057/9781137436511.0010

In the Anglesey case, the efficiency of the process was significantly impeded by CCHQ email accounts having inadequate data storage capacities to cope with the sign-off traffic and file transfer, especially at the times when large numbers of candidates sent digital proofs simultaneously. In Anglesey, the process described above was delayed by, at least, one week because emails containing digital leaflet proofs for sign-off were not being received by the intended recipients in the sign-off hierarchy. Smith (2011) describes the bureaucratic culture of the Scottish Conservatives as 'banal activism' in which 'discursive artefacts' like leaflets and press releases provided a distraction for activists from the reality of the electoral 'crisis' of the Conservative Party in Scotland. Similarly, the checks-and-balances in place to monitor traditional campaign literature, and the bombardment of digital internal communications in the party in 2010, were so cumbersome a process that candidates without significant administrative support could become buried beneath processes and bureaucratic distractions. Ultimately, these distractions significantly limited temporal resources and detracted from the face-to-face campaign out-and-about in the constituency.

Even though, at times, the limitations of CCHQ's email systems led the process to collapse, in 2010, without the use of email whatsoever the bureaucratic intensity of the sign-off process would have been impractical and relatively unworkable. Email as a medium had been embraced by Cameron's Conservatives in a manner in which the medium was assimilated into the party's hierarchal structure. Therefore, as an organizational tool, email facilitated the party's hierarchical tendencies in the offline world without the need for face-to-face interaction or oral communication among the individuals in the chain of command. The intensively cautious approach taken towards the release of paper based printed media into the public domain was not applied with the same scrutiny to candidates' online publications like blogs and social media (Fisher et al. 2011). Therefore, this would suggest that Cameron's Conservatives considered traditional media to be of greater importance to their overall election objectives than new media, but that some internet technologies like email had become essential organizational tools in order for the party to effectively facilitate the scrutiny of its printed media in a hierarchical fashion.

DOI: 10.1057/9781137436511.0010

PPC for Ynys Môn and social media

In my experience, candidates had relative freedom and autonomy in terms of their campaign's internet presence in the respect that, other than the brief guidance in the campaign pack, there were no formal written rules or procedures distributed by the central party to Conservative candidates. This is supported further by responses given by Respondent Four. He claims that he 'offered to write a candidates guide on the best practices for the use of blogs and Twitter. The party initially thought it was a good idea, but nothing happened'. The respondent believed that 'candidates seemed too afraid to actively use blogs and Twitter. They were often worried they would say something wrong and there would be consequences if they deviated from the party line'. The respondent testifies that he did not witness any attempts to manage the Conservative blogosphere by CCHQ London or communications professionals like Andy Coulson. In keeping with Kavanagh (2013), Respondent Four believes that 'Labour did try to manage their blogging community. But, as you would expect, they did it with the typical top-down approach'. This typifies a Conservative belief that the Tories took a different approach to the management of new media in GE2010 when compared to Labour.

Personally, I invested little time in blogs and social media during the Anglesey campaign when compared to other campaign methods. Nevertheless, as a relatively young candidate, I felt it important to invest at least a minimal amount of time in social media in order to make potential political contact with, what I assumed would be, younger voters. I carefully controlled my Facebook profile privacy settings and consciously sanitized my online postings. For security reasons, I chose to limit what could be viewed and interacted with on my Facebook profile, and developed a more publically viewable and interactive 'Facebook Page'. Therefore, I existed on Facebook as one individual with two linked, but ultimately separate, digital personas – the personal and the political. The second page was listed as a politician page on Facebook, which gave Facebook users the option to search my name and follow my campaign in their News Feeds by clicking 'Like' on my page. Subsequently, the user became linked in interactive terms to my page. The number of users that 'liked' the page accumulated over time. I observed also a number of fellow candidates using the page function in Facebook during their campaigns (Chapter 4).

DOI: 10.1057/9781137436511.0010

In terms of the Anglesey campaign, in a digitally interfaced manner, I integrated a simple blog application on my personal website (Ridge-Newman 2010a) in conjunction with my pages on Facebook (Ridge-Newman 2010b) and Twitter (@RidgeNewman 2010). Using an online application, I linked my political Facebook page to my Twitter page in order to duplicate my Facebook postings automatically onto my Twitter feed. I had taken the time to compare both my Facebook fans and my 'Followers' on Twitter. I noted that the audiences of each comprised of significantly different individuals. Therefore, the duplication of my Facebook postings on Twitter expanded my audience by almost double. This enabled me to save the amount of time I spent using internet applications in order to focus my time on the offline campaign. Facebook, and subsequently Twitter, acted like a notice board for the advertisement of my blog posts. Most blog posts were created in order to communicate the work I was doing in the constituency and to publish my views on carefully selected local and national issues. In accord with Respondent Four's testimony above, I noted this to be in keeping with the majority of fellow social media using candidate colleagues. I held the distinct belief that these postings would not lead to significantly greater numbers of votes, but rather motivate my supporters and campaign team. The use of social media in the Anglesey campaign was primarily a tool for activist mobilization in keeping with Gibson's (2013) grassroots model of the 'citizen-initiated' campaign.

PPC for Ynys Môn and email

Email was used regularly within the campaign team itself in order to organize canvass teams and locations. My campaign coordinator and agent shared a Gmail email account with me, anthonyforynysmon@gmail.com, in order to organize the internal and external campaign team contacts more effectively than the functions provided by my public email on Mail.com, ant@politician.com. The Gmail account had a calendar function which was used by the campaign coordinator to input my daily appointments and weekly campaign/canvassing schedule. Email in the Anglesey Conservatives' campaign was used primarily as an internal communication tool rather than a marketing device. Information that included instructions for campaign days was regularly emailed to the appropriate supporters and members of the campaign team. This

DOI: 10.1057/9781137436511.0010

process worked effectively and had greater impact than social media. The campaign team's ages ranged between 20 and 70, with the majority being around 60 years old. As the campaign progressed, email seemed to become a more established and effective medium for communication in the older cohorts of Anglesey Conservatives.

I made requests to the same individuals to collect constituents' email addresses while canvassing. However, the Anglesey based campaign team generally neglected to ask electors for their email addresses. The common response to my request from members of the campaign team was that they did not understand why the collection of email addresses was necessary. After repeated explanations of why I believed it to be important to collect as many voters' emails addresses as possible during the campaign – for direct marketing and campaigning purposes – the activists continued to neglect asking for email addresses on the doorstep. Instead, they held a distinct preference for canvassing in line with their experience in previous campaigns. For the Anglesey Conservatives, this meant not collecting traditional canvass data, but instead distributing leaflets and getting the candidate to 'shake hands with as many voters as possible' before election day. This shows how changing campaigning activities within the party at the constituency level can exhibit a form of inertia and resistance to change if the experienced individuals of the campaign team believe the changes to be unnecessary.

PPC for Ynys Môn and campaign websites

The political website that seemed to make the most notable impact on local and regional debates concerning Anglesey was the anonymous blog 'The Druid'. The Druid later 'came-out' and revealed himself to be Paul Williams (Williams 2010b), a member of the Anglesey Conservatives campaign team, who later went on to represent the Conservatives as their Welsh Assembly candidate for Ynys Môn in May 2011. The advent of The Druid blog in January 2010 marked the arrival of some form of e-politics on this remote and rural island in North West Wales. The blog, which often posted gossipy content, quickly became the centre of local gossip itself, because many Anglesey activists with or without an internet connection seemed curious about the identity of the island's very own cyber activist. However, the blog seemed more of an extension of Anglesey's often bitter social and political 'backbiting' (Hughes 2011).

DOI: 10.1057/9781137436511.0010

Ergo, the blog symbolized rather more the gossip culture of an island community than being a political tour de force symbolizing the arrival of e-democracy on the island. One case that somewhat illustrates the Anglesonian political backbiting characteristic is an apparent attempt to capitalize on The Druid's reveal with an attack on the Conservative Party by the incumbent Labour MP for Ynys Môn (*Daily Post* 2010).

As discussed in Chapter 4, arguably blogs may have been a natural niche for Conservative activists in the run-up to GE2010. However, such modes of activism were not always helpful to the party's collective aims and image. This is especially apparent with trends in anonymous blogging, like that seen in the cases of Tory Bear and The Druid, which appear to have been strategically designed for a great public reveal in order to grab media coverage and wider attention. In general, these were innovative strategies that benefitted the individual rather than the party collective.

Personally, I managed a very primitive form of a candidate's blog while campaigning for the general election. I wrote a number of letters to the *Holyhead and Anglesey Mail*, and posted the text of any letters written into a webpage on my personal campaign website. I posted, usually after the event, notifications of where I had been on the island and details of my campaigning activity with the aim of informing constituents of my activities. My personal campaign website averaged 62.5 hits per day during my 117 days on the campaign as candidate, with an approximate range of variation between 32–94 hits per day. The peaks and troughs of website hits appeared to have some correlation with campaign events, like major deliveries of literature and my appearances on radio or television; and periods when my profile as a candidate was less publically visible. Using an online tracking facility, I was able to identify the general geographical locations from which my website was being accessed. Less than 10 per cent of the hits were from the North Wales region. The majority were from within the UK, but there were a significant number of hits from worldwide locations. Therefore, the website's impact on local electors could have been only minimal.

This is further supported by one occasion, during the final week of the campaign, when I spoke to a constituent on his doorstep. He had received my election literature through the Royal Mail postal service, but complained that there was not enough personal information about me on my election leaflets. I explained that the address for my campaign website was printed on the leaflet and that there was more information about me on the website. However, he replied saying that he should not have to go to a website and that he should have received the information about me in leaflet form. Although this was an isolated case, it is typical of the kind of disregard for internet technology, especially in terms of political campaigning, which I found amid the local Anglesey electorate in 2010. Therefore, in terms of campaign impact on the Anglesey electorate, the 11.5 per cent increase in the vote share achieved by the Conservatives on Ynys Môn (Dods 2010: 534), 7.7 per cent above the national trend (BBC News 2010), was most likely because of the traditional campaigning techniques used like door-to-door canvassing and printed literature, and a highly visible and energetic campaign presence across the constituency.

Albert Owen, the island's MP, had a Labour Party page (Owen 2010), but did not have a live website until the final weeks of the campaign. His website was basic in terms of its content, features and aesthetic. Ynys Môn was also the Welsh Assembly seat of Ieuan Wyn Jones AM, the then incumbent leader of Plaid Cymru. Therefore, like Labour, Plaid Cymru had a buoyant and established support base and activist presence across the island. Dylan Rees, the Plaid Cymru PPC in 2010, required an extra 1243 votes in order to win the seat from Labour's Albert Owen. Anglesey was the number three target seat for Plaid Cymru in Wales. Plaid Cymru invested visibly significant resources into their campaign. Rees had the most sophisticated website (Rees 2010) of all the main party candidates on Anglesey. However, on election day, the Plaid Cymru vote share decreased by 4.9 per cent, and although the Labour candidate was returned as the island's MP his vote share decreased also by 1.2 per cent (Dods 2010: 534). Therefore, there was no observable correlation between electoral success and investment in campaign websites on Anglesey in 2010. This finding is in keeping with Andy Williamson's (2010) assessment of GE2010 from which he concludes that, although the internet was an integral bureaucratic and organizational tool, web-based campaigning activity in 2010 was insufficient for it to be deemed an 'internet election'. Therefore, Anglesey's lacklustre e-campaign seems in keeping with wider trends in local campaigning during GE2010 (Fisher et al. 2011).

DOI: 10.1057/9781137436511.0010

8
Cyber Toryism

Abstract: *Chapter 8 is used to develop the case for Cyber Toryism as a cultural singularity in the history of the Conservative Party in existence mainly between 2008 and 2010. The chapter concludes that generally the internet acted like a lubricant oiling Conservative Party processes, which in turn resulted in greater fluidity within networks and organizational and campaign operations. It is argued that this loosening of centralized control allowed for shifting power dynamics and subsequent culture change to occur. It is found that the younger cohorts were central to this change in party culture, which had remained more traditional since John Major's leadership; and that the advent of David Cameron's leadership acted to punctuate, in other words speed-up, technocultural evolutions in the life of the wider party.*

Keywords: Culture change; Cyber Toryism; network society; organizational culture; party decentralization; party organization

Ridge-Newman, Anthony. *Cameron's Conservatives and the Internet: Change, Culture and Cyber Toryism.* Basingstoke: Palgrave Macmillan, 2014. DOI: 10.1057/9781137436511.0011.

In chapters 2–7, themes have emerged that suggest that the Conservative Party has undergone some organizational adaptation to the advent of internet-based technologies. This appears to be an ongoing evolution in progress (Howard 2006; 2010). As Eran Fisher (2010) might argue, this evolution is symbolic of an increasingly globalized age in which new technologies facilitate a different societal environment. In terms of the impact of the internet in the Conservative Party, the most significant themes seeming to characterize the party's organizational culture in the run-up to GE2010 and beyond include control and transparency; digital and age divides; digital bureaucracies; digital campaign enhancement; dissolution of geographical boundaries; heterogeneity of Conservative associations; integration of new technologies in traditional hierarchies; new Cyber Tory elites; and rapport and digital trust signals. These themes appear to have been manifested in a largely dichotomous manner in the party's organizational culture, which in turn means that the party has exhibited a range of different characteristics within subcultures.

The digital divide is mostly associated with generational demographics (Chapter 6). The older demographic group of the party generally maintained a commitment to traditional organization and campaign techniques. The younger demographic group of the party, which was largely represented by participants in the networks of CF and the Conservative blogosphere, engaged in complementary activities and behaviours in both the on- and off- line worlds. These individuals and collectives both supported Conservative Party campaigns and challenged the status quo of the party's organizational culture.

There were also significant geographical variations dependent on the specifics of the seat location and the nature of the local association, candidates and campaign factors (chapters 5, 6 and 7). For example, the more metropolitan Conservative cohorts in London and Surrey, where high speed broadband was more prevalent, generally used the internet in a more established, prominent and social manner in campaigning and party organization when compared to the more remote communities like Anglesey in North West Wales. Furthermore, key target seat participants seem more likely to have had a greater internet presence and intensity of interface with internet technologies than non-target seat participants.

When the themes of this book are integrated with existing literature it is clear that the Conservatives were not the only British party undergoing an adaptation to the use of internet technologies. However, current research does indeed suggest that the Conservative Party's

DOI: 10.1057/9781137436511.0011

specific response to the dynamics of the new media environment and its subsequent intraparty cyber culture was in some ways unique when compared to the other major British parties (Lilleker and Jackson 2010). The Conservatives certainly changed their approach to the internet from 2006 when compared to earlier periods. As the party embraced internet technology more, it would seem that it synthesized with some of its historic characteristics like autonomy, pragmatism and the importance of party leadership (Charmley 1996; Taylor 2008) that led to new cultural outcomes. This concluding chapter further expands on the main themes that have grown out of this book and places these in the context of the argument for Cyber Toryism as an observable technological subculture within the Conservative Party's wider organizational culture.

Digital and age divide

The two main national groups of collective participants in the official Conservative Party organization – (1) Conservative associations and (2) CF branches – are primarily divided by age. The age divide, in addition to secondary factors like the differences in the ways in which these two main groups are structured and organized, led to different responses to the use of internet technologies in the run-up to GE2010. Both groups are social in nature and, in addition to the party's wider organization, have evolved significantly over the party's history (Ramsden 1995; 1996). Between 2006 and 2012, CF participants generally embraced the uses of online social networking tools with greater enthusiasm than older participants and used them as tools for socializing within a campaign motivated context. This helped dissolve the confines of the traditional geographical limitations under which Conservative associations have generally operated in the past (Whiteley et al. 2002). The use of the Facebook application by younger members of the party allowed for greater networking activity both on- and off- line.

There became two distinct organizational cultures within the party in the run-up to GE2010. Amid the general heterogeneity of Conservative associations nationally, there were those collectives of individuals who engaged primarily and sometimes exclusively in the traditional communication and structural processes of Conservative Party organization; and those who transferred and assimilated much of their activity within the organization of the Conservative Party to the internet – with the

DOI: 10.1057/9781137436511.0011

development of a mutually beneficial relationship between on- and off-line Conservative Party operations. However, as Gibson and McAllister (2013) suggest, cyber organization that neglects entirely traditional offline party approaches is somewhat impaired when it comes to practical impact in the offline world. As a new intraparty cyber subculture grew, there was a proliferation of dynamic interactions between both collectives and individuals internally that led to a distinctly alternative organizational behaviour and culture when compared to the more traditional view of Conservative Party interaction and organization (Bale 2012).

Philip Howard (2006; 2010) asserts that, firstly, technology can evolve and, secondly, it has the power to impact on individuals and groups. In this case, certain technocultural evolutions in wider society appear to have led to new media innovations that impacted at micro- and macro-cultural levels in the Conservative Party. It would appear that the primary factor influencing the behavioural and cultural change was the role of internet applications, primarily, between 2008 and 2010 at the grassroots level. Observations suggest that by 2012 the use of social media and other internet technologies was less exclusive to the younger groups within the party and that the internet's place in the culture of the party had become significantly more established in the wake of GE2010. It seems pertinent to designate this phenomenon as a distinct cultural singularity within the organizational culture of the British Conservative Party during a certain period in its history. Therefore, this Cyber Toryism, as I call it, is applied exclusively to the run-up to the GE2010 period in order to distinguish it as a subculture from the party's traditional and overarching meta-culture at other times. The evidence presented in the book suggests that since GE2010 Cyber Toryism might have been assimilated into a modernization of the party to a point that is becoming more of a cultural norm in the daily behaviour and practice of party participants. This would suggest that applications like social media that were thought of new media not that long ago have already reached a stage of maturation in which they are more firmly established as part of the party's culture.

In 2010, the quid pro quo attitude and technologically elevated status of the younger, more dynamic and less geographically constrained Cyber Tory participants meant that they were of value to the party as a campaigning resource. However, they took also a less deferential approach to the party hierarchy in terms of maintaining the cultural status quo in their participation when compared to their more traditional counterparts. In the run-up to GE2010, internet applications gave the ordinary Conservative participant a greater and more independent voice while simultaneously allowing the individual, and/or collective group the ability to support the party in traditional campaign and organization roles. It allowed for also individual innovation and leadership to shape culture change (Schein 2010) through the use of alternative digital tools that were further empowered by advances in internet technology and uses. This had impact in both the on- and off- line culture of the party in different ways because of the multi-dimensionality of such participation (Cantijoch et al. 2011).

Although popularity, in terms of 'Fans' and/or 'Followers', in social networks demonstrated the importance of the likeability of the party leadership, Cyber Toryism, as a culture, seems to have partly embodied a further erosion of traditional party deference and, thus, constitutes a shift away from the historic hierarchical control of the central party's operations. It would seem that the party's active engagement in the internet ultimately impacted on the central party's traditional control over these operations and output. Therefore, the central party's tight professionalized grip, which was strengthened between the early 1960s and 2000s, with a progressive professionalization of the party's operations and communications, was seemingly involuntarily loosened in the late 2000s through a culture of mass online engagement at the Conservative Party grassroots in the form of the generally organic evolution of Cyber Toryism.

The Conservative Party's relatively natural cyber evolution is further highlighted when placed in the context of the Labour Party, which, in comparison, has been found to have used the internet to take a more top-down approach to generating grassroots activism (Straw 2010). Therefore, any technocultural impact on Labour Party organization could be deemed more deliberate. Moreover, the advent of Cyber Toryism in the intraparty dynamics of the Conservatives meant that the widespread and pragmatic use of the internet at the party grassroots acted like a lubricant, which oiled party operations and campaigns, and in turn gave a greater fluidity and mobility to individual participants and organizational mechanisms (Lin 2004) and, therefore, acted more as a catalyst for change than a tool for change.

Catalyzing Cyber Toryism

Through a process of political socialization (Kavanagh 1972) and incentivized campaign participation, individuals learnt to synthesize their

DOI: 10.1057/9781137436511.0011

online activities with Conservative activism. Between 2006 and 2010, the party's relationship with the internet was most saliently symbolized in WebCameron, which appears to have acted as a catalyst that in turn contributed to a stimulation of an organic evolution of innovation in social media uses at the party's grassroots. This phenomenon seemingly proliferated through a culture of learning, copying and adapting to the uses of internet applications. Therefore, it seems appropriate to identify WebCameron as the symbolic beginning of Cyber Toryism for the Conservative Party as a whole.

Furthermore, it seems plausible to suggest that the visible use of internet technologies and applications by prominent Conservative leaders, like Cameron and Johnson, in a campaign context, was a signal to those within the party organization, who were actively engaged in the use of the internet in both personal and political capacities, that innovative use of the internet for the gain of Tory Party operations was appropriate (Schein 2010). It would seem that this latent but symbolic message from the party leadership acted as an unwitting catalyst for subsequent growth in the use, experimentation and innovation of internet applications that assisted in the organization of intraparty affairs in the run-up to GE2010. Applications like Facebook acted as online venues for party participants to congregate and engage in the organization of Conservative Party events, discussions, debates and campaigns. This in turn led to generally well-organized and well-attended party events and operations in the offline world. Whether these modes of on- and off- line participation best fit differentiation, replication, integration or independence hypotheses remains somewhat unclear, but other work does suggest that political culture may be exhibiting an new type of cyber-facilitated behaviour in general (Gibson and Cantijoch 2013). Therefore, it is important to note that although the phenomenon of Cyber Toryism is unique to the specific case of the Conservative Party narrative, similar phenomena may have been observable in other parties in the run-up to GE2010.

Although the cultural dynamics of Cyber Toryism are likely to have been punctuated due to visible endorsements of the use of internet technologies by influential figures at the top of the party hierarchy, which itself was a result of wider technocultural evolutions in wider society, it is evident that the subsequent evolution of the phenomenon perpetuated firmly from within the ranks of the party's grassroots. Prominent Tory activists, candidates and lobbyists, like Richard Jackson, Iain Dale and Tim Montgomerie, respectively, took innovative steps to embrace the

DOI: 10.1057/9781137436511.0011

first uses of Facebook and blogs in effective ways, which in turn helped to enthuse a new generation of Conservatives for whom the internet was already part of their daily lives. Although, at times, appearing somewhat masked by the blurring of traditional boundaries (Youngs 2009), the proliferation of Cyber Toryism developed with the transfer of the behaviours and practices of the traditional political culture of the younger members of the Conservative Party on to a new mode of internet-based participation within the party (Gibson 2013). It would seem this resulted in a coexistence of both on- and off- line cultural phenomena, which were observable and learnable by other Cyber Tory neophytes. A culture of copying, adapting and enacting the use of new media in Conservative Party organization and campaigns meant that the phenomenological significance of Cyber Toryism continued its growth and proliferation from a London centric base, prior to and during the 2008 elections, to wider national reaches in the run-up to GE2010.

Seemingly, by 2012, the veil of latency of Cyber Toryism had begun to lift from the party's consciousness. Although the party did not use the tag 'Cyber Toryism' to describe the observable behavioural changes in its organization, the identification of a cultural change had begun nevertheless to receive some cognitive analysis by members of the party's sophisticate and grassroots participants. This realization and enlightenment in the party's consciousness of the role of social media as a powerful force for change within the Conservative Party further impacted in the dialogue, discourses and agenda of party debates, operations and campaign literature.

In 2010, there was little centralized best practice or guidance on the use of internet-based media in Conservative campaigns and organization. Whereas, by 2012, the central party operations had submitted to the arrival of widespread usage of internet applications by individual participants in the party in the form of publishing an extensive guide on e-campaigning that was largely set apart from literature and guidance pertaining to traditional media and campaign techniques. This was a step towards assuming some central control of the relatively uncontrollable nature of social media.

Organizational change

Through the widespread uptake of Cyber Toryism at the grassroots by 2010, the role of social media had impacted on the traditional structure

DOI: 10.1057/9781137436511.0011

and organization of the Conservative Party in the form of shifting some of the traditionally centralized and controlled power, over party communications, message and operations, to Conservative participants at the grassroots for whom the traditional structure of centralized control through hierarchy and deference had previously limited their engagement (Ball 1994b; Seldon and Ball 1994). Therefore, the advent of the internet and internet-based applications, like social media, impacted on the organization of the Conservative Party in providing a new environment and social tools with which users were able to claim greater control of party functions through innovation and imitation.

The lack of top-down control and guidance on matters relating to the internet meant that the traditional autonomy of the association (Ball 1994a) was extant in terms of their approach to the use of new technology and media. Therefore, each local association took an autonomous approach to the internet. Some associations, like Richmond Park, highly embraced technological change and young people in respect to their campaign operations. Through the extensive use of online technologies, like Facebook groups, those campaigns which sourced young activists, through social media, demonstrated a significantly greater engagement and participation in specific campaigns than those which took a more traditional approach. In the successful cases, Facebook and email were used actively as tools to sell participation in campaigns. These internet technologies were used to sell incentives to potential activists, like drinks and refreshments, in order to encourage participants to take the step from weak online interactions (Margetts 2006) to full offline active campaign support. Young people campaigned less for party deference, loyalty and tribal allegiances and more on a quid pro quo basis.

Facebook specifically helped remove the traditional geographical constraints over communication in the way in which traditional associations operated. Therefore, Facebook acted as a virtual association/participant which oiled organizational processes thus making party operations more fluid and bringing closer the national party into more intimate spaces within online venues. Facebook was also, for many associations and political figures, a shop front or window in which they were able to display a national gauge of popularity through the numbers of fans and supporters they had in any given Facebook page. Facebook's prominence is perhaps demonstrated mostly in the perception of one Conservative insider that, for a newly selected candidate, a presence on Facebook had become a priority in line with the traditional press release as a mode to announce selection. Facebook was also a political leveller in that for the first time the leadership of a party and a candidate were presented in a medium on a more equal playing field. In turn, Facebook helped facilitate a technology centred culture at the party's grassroots that had not been a constituent part of previous elections.

The resultant impact on the party's traditional constitution was that the en masse use of these media at the party's grassroots shifted some of the power previously held by a few individuals in the party's elite centres of power, like CCHQ and the party leadership, to a wider collective of Conservative participants at lower levels of the party's traditional hierarchical structure. There were two main manifestations of this in the party's organization. Firstly, the use of internet applications oiled the party's operational processes meaning that aspects of the party's organization, like mobilizing campaign activism among young people, was more fluid in its execution than it would have been otherwise (Gibson 2013). Secondly, the observed widespread Cyber Toryism elevated the impact of blogs like ConservativeHome and Iain Dale's Diary, thus giving some participants at the party's grassroots a prominence and power they would have not had otherwise and creating new Cyber Tory elites in the process (Hindman 2009).

It seems appropriate to suggest that this in turn helped loosen the grip that the central elites had over what was generally and traditionally viewed in the public sphere as Conservative Party controlled output and internal affairs. It seems both existing and new elites acted to help catalyse the proliferation of Cyber Toryism at the grassroots. Therefore, it is plausible to suggest that the advent of the internet and its subsequent applications played roles which have resulted in a general loosening of the more CCHQ centric communications culture in which the party had progressively tightened since the 1960s. In turn, the loosening of traditional party organization to a more cyber-influenced organization has both strengthened existing and new elites in some cases, while allowing a democratization of the Conservative Party's grassroots engagement in other respects. This is especially the case for Facebook use which has facilitated a more fluid and networked grassroots organization (Gibson and Ward 2012).

Those technologies which grew from organic uses at the grassroots upward, like blogs and social media, had a greater impact on the wider party's organizational culture than those technologies, like MyConservatives and MERLIN, which were developed and managed by

DOI: 10.1057/9781137436511.0011

central party operations. Therefore, it seems plausible to suggest that it was the behaviour of younger participants at the party's grassroots, in the run-up to GE2010, which made the most significant impact on the party's organizational culture in terms of technological use, behaviour and innovation. As the history of these events unfolded in the dynamic environment of a high profile long campaign, the central party was engaged in observing, understanding and learning these changes, and was thus a step behind those engaged in the real-time participation in Cyber Toryism at the grassroots. Ultimately, contemporary Conservative Party culture appears to have morphed from a traditional organization characteristic of modernity to a complex soup of on- and off- line networked interactions in which the new found fluidity and organizational looseness facilitates dynamic shifting power relationships between the party elites, new elites and grassroots in a non-predetermined collection of diverse trajectories that has impacted on political individuals, groups, institutions and systems.

Control and transparency

The use of internet technologies impacted on the party's organizational culture significantly less in processes over which the central party maintained greater traditional control, like for example WebCameron. The process of candidate selection is another example of a process in which CCHQ and association cadres maintained their traditional control in terms of the process's execution. However, candidate selection is also an example of how members of the party's affiliated blogging community broke the party's traditional protocol in releasing otherwise secret internal affairs into the public domain. This forced the party into a position of revealing more of its inner workings than it would have traditionally. Therefore, in this case, the advent of the use of specific internet applications led to a greater transparency in central party operations. Furthermore, the visible and public nature of blogs and social media meant that any acts by party participants could be viewed by individuals and collectives both inside and outside the Conservative Party. This behaviour, which to some could have been believed to be impertinent and defiant, and a demonstration of a rejection of the party's traditional deferential behaviour, was viewed by some inside observers as a challenge to Conservative Party unity and the party's official messages.

DOI: 10.1057/9781137436511.0011

ConservativeHome's development, as a technocultural symbol of a globalized network society and phenomenon within the party's organizational culture, was less like the organic evolution of the technological uses observed in other internet applications, like social media and other Conservative blogs. Therefore, it is plausible to suggest that Cyber Toryism was coevolutionary in its expansion. Montgomerie's early objective was to use the blog to shift power from the central party to the grassroots. Although some power was distributed to ConservativeHome participants at the grassroots of the Conservative Party, much of the power that was yielded by the blog application was channelled into the hands of Montgomerie himself, thus elevating him to a position of elite status in the party, and beyond, in his own right. In this case, the power and the impact of the internet is most apparent, where the innovative use of internet applications through the vision of one individual was able to challenge Conservative Party hierarchy through that individual's rise in prominence in the party's unofficial ranks (Lilleker and Jackson 2010).

The manifestation of the unofficial prominence of specific individuals in the party through the medium of the internet is in itself a new phenomenon to which the Conservative Party is yet to adapt entirely (Gibson et al. 2013). The Montgomerie case is an outlier when his extraordinary achievement is compared with the collective mass of Cyber Toryism elsewhere in the party. However, in view of the evolving, observing, learning, copying and adapting cyber culture at the grassroots, the party could indeed see a number of Montgomerie types grow out of the future of Cyber Toryism – unless the party adapts significantly its organization to regain the tight central control that began to be eroded in the run-up to GE2010.

Conclusions

The narrative of Cyber Toryism in the run-up to GE2010 is one that represents a changing party organization and wider political culture. Moreover, it represents and tells us something about the changing nature of British society and beyond (Thakur 2006). Britain is in a transitional phase. It seems to be culturally shifting from modernity to what some call the globalized network society. In this case study of the Conservative Party, we see evidence of these two cultures coming together and in some respects clashing and creating cultural tensions (Clarke and Kaiser 2003). The historic cultural traditions of the Conservative Party have

DOI: 10.1057/9781137436511.0011

continuously evolved over hundreds of years. In the run-up to GE2010, wider technocultural trends impacted on the party's organization with an intensity that had not been experienced since the 1950s and 60s amid the advent of political television. The party's technocultural evolution in the mid to late 2000s was rather more punctuated. Preceded by a period of relatively incremental adaptations, from 2006, under the new leadership of David Cameron, rapid and significant change occurred over a shorter period. For over half a century, the Conservative central elite had tightened increasingly its grip of control over the wider party organization. By 2010, with the advent of networked internet technologies, something significant had changed.

The nature of the internet means that it allows for unilateral approaches to be pursued at lower cost and with greater ease. In 2010, this in turn created a greater degree of heterogeneity and autonomy in some Conservative associations, which was rooted in the choices, culture and leadership of the local associations. In earlier chapters, there are a number of examples presented of technologically centred innovations, which contributed to a rich diversity embodied in Cyber Toryism. The collective impact of diverse unilateralism and innovation in the use of internet technologies in the Tory Party seems to have assisted the loosening of centralized power structures, which in turn shifted increased organizational power to the party grassroots. The resultant impact of this was to create, in rare cases, new cyber elites. These Cyber Tory leaders acted as catalyst for further change in the ways in which individuals collectively behaved and interacted in the Cyber Tory environment.

As Cyber Toryism becomes more normalized behaviour in the party's organizational culture, the party risks disenfranchising party participants who are on the periphery of the new technocultural practice. There remains potential for further digitization and empowerment of the party's grassroots. The central party could use this to its advantage if it were to orientate its mindset in order to capitalize on newer forms of activism. For example, internet technologies hold the potential for activists to work together in diverse remote locations. Less geographically tethered e-campaigns could be focused into winning key marginal seats from anywhere in the country. Internet use is also more cost effective for activists and would allow individuals to do their bit without needing to travel great distances and without the investment of great resources. Furthermore, if the party can successfully synthesize its aging supporters

DOI: 10.1057/9781137436511.0011

with its internet strategies, participants with mobility difficulties could provide an army of home based cyberactivists in the future.

However, as discussed in Chapter 7, currently, cyber interactions are not as effective as face-to-face interactions in terms of maintaining camaraderie and motivation. Like Anglesey in 2010, there are parts of the UK which are yet to benefit from satisfactory broadband links. Furthermore, these globally accessible activists would be a challenge to campaign authenticity. But if the Conservatives, and indeed other parties, are able to further innovate in order to overcome such challenges, or if the wider political culture changes in a manner that reduces the capacity of such challenges to limit the campaigns, then the power of the internet could be harnessed to further dissolve geographical barriers and empower activists across regions to apply support to the most critical, election determining, seats.

DOI: 10.1057/9781137436511.0011

Bibliography

Primary sources

Blog and news articles

BBC News (2009) 'McCanns' anger at Tory activist', *BBC News*, 9 January 2009, online: http://news.bbc.co.uk/ (home page), date accessed 12 January 2009.

BBC News (2010) 'Election 2010: National Results after 650 of 650', *BBC News*, online: http://news.bbc.co.uk/1/shared/election2010/results/, date accessed 22 May 2010.

Chivers, T. (2009) 'New Tory website MyConservatives.com collapses on launch', *The Telegraph*, 2 October 2009, online: http://www.telegraph.co.uk/ (home page), date accessed 25 April 2014.

Cole, H. (2009) 'An apology from Matt Lewis', *Tory Bear*, 9 January 2009, online: http://www.torybear.com (home page), date accessed 12 January 2009.

Daily Post (2010) 'Anglesey blogger The Druid reveals himself', *The Daily Post*, 20 November 2010, online: http://www.dailypost.co.uk/ (home page), accessed 16 December 2011.

Forsyth, J. (2010) 'Labour sack candidate caught up in Twitter scandal, *The Spectator*, 9 April 2010, online: http://blogs.spectator.co.uk/ (blogger page), date accessed 9 May 2014.

Fryer, J. (2014) 'Sex, spies... and the truth about my hair: Cameron sacked Michael Fabricant as Tory vice-chairman for being so indiscreet. But has he learnt

DOI: 10.1057/9781137436511.0012

his lesson? Not on your life!', *Daily Mail*, 10 May 2014, online: http://www.dailymail.co.uk/ (home page), date accessed 17 June 2014.

Hannan, D. (2009) 'My speech to Gordon Brown goes viral', *The Telegraph*, 21 March 2009, online: http://blogs.telegraph.co.uk/ (blogger page), date accessed 22 July 2010.

Hern, A. (2013) 'Conservatives remove WebCameron from YouTube', *The Guardian*, 14 November 2014, online: http://www.theguardian.com/ (home page), date accessed 25 April 2014.

Hodgkinson, T. (2008) 'With friends like these ...', *The Guardian*, 14 January 2008, online: http://www.theguardian.com/ (home page), date accessed 15 May 2014.

Hughes, O. (2011) 'Anglesey councillors warned local authority could be made to merge', *The Daily Post*, 18 March 2011, online: http://www.dailypost.co.uk/ (home page), date accessed 23 June 2014.

Jones-Evans, D. (2009) 'Anglesey suffers more economic woe', *Western Mail*, 3 October 2009, online: http://www.walesonline.co.uk/ (home page), date accessed 12 December 2012.

Payne, S. (2013) 'Farewell WebCameron, and the legacy of Steve Hilton', *The Spectator*, 15 November 2013, online: http://blogs.spectator.co.uk/ (blogger page), date accessed 25 April 2014.

Salman, S. (2011) 'Is Cameron's "big society" reserved for the rich?', *The Guardian*, 18 May 2011, online: http://www.guardian.co.uk/ (home page), date accessed 12 December 2012.

Simpson, J. (2014) 'Tory candidate for Brentwood South resigns after tweeting that Islam was the "religion of rape"', *The Independent*, 4 May 2014, online: http://www.independent.co.uk/ (home page), date accessed 10 May 2014.

Urwin, R. (2014) 'Absolutely Fabricant: the sacked Tory vice-chair who's whipping up a Twitterstorm', *London Evening Standard*, 11 April 2014, online: http://www.standard.co.uk/ (home page), date accessed 17 June 2014.

Williams, M. (2010a) 'Anglesey businesses urged to get connected', *The Daily Post*, 15 December 2010, online: http://www.dailypost.co.uk/ (home page), date accessed 4 April 2012.

Williams, P. (2010b) 'The Druid Revealed', *The Druid*, 19 November 2010, online: http://druidsrevenge.blogspot.co.uk/2010/11/druid-revealed.html, date accessed 11 November 2011.

Woodward, W. (2006) 'Tories Unveil Their Secret Weapon: WebCameron', *The Guardian*, 30 September 2006, online: http://www.guardian.co.uk/ (home page), date accessed 23 July 2012.

DOI: 10.1057/9781137436511.0012

Wright, O. (2014) 'Michael Fabricant: Llamas, incest, bestiality, and the demise of the Tories' Deputy Chairman', *The Independent*, 21 June 2014, online: http://www.independent.co.uk/ (home page), date accessed 17 June 2014.

Emails and letters

CCHQ London (2010) 'E-campaigning in an election', Conservative Campaign Headquarters London, two page General Election Memorandum 2010.

Fabricant, M. (2014) Email from Michael Fabricant MP to Anthony Ridge-Newman, 24 June 2014.

Maude, F. (2006) Conservative Party Chairman Welcome Letter to A. Ridge-Newman, 2 November 2006.

Pickles, E. (2010) Conservative Party Chairman Welcome Letter to A. Ridge-Newman, 25 March 2010.

Ridge-Newman, A. (2007) Email reply to a Spelthorne Conservatives officer, via ant@politician.com, 4 November 2007.

Ridge-Newman, A. (2008a) Email sent to nine Surrey Conservative officials, via ant@politician.com, 15 September 2008.

Ridge-Newman, A. (2008b) Email sent to nine Surrey Conservative officials, via ant@politician.com, 23 September 2008.

Interviews

Respondent One: Harwich Town councillor; key national CF activist; National Chairman of CF; Conservative PPC for Bath, interviewed 28 March 2011.

Respondent Two: Deputy Head of New Media at CCHQ, interviewed 23 May 2011.

Respondent Three: Runnymede Borough councillor; chairman of Runnymede and Weybridge Conservative Association, interviewed 29 March 2011.

Respondent Four: Prominent Conservative blogger; LBC radio presenter; campaign coordinator; Conservative PPC for North Norfolk, interviewed 3 May 2011.

Respondent Five: Co-editor of the ConservativeHome blog; *Telegraph* newspaper journalist, interviewed 31 March 2011.

Respondent Six: Key London CF activist; National Chairman of CF; *Sun* newspaper journalist, interviewed 31 March 2011.

DOI: 10.1057/9781137436511.0012

Respondent Seven: Councillor and leader of the Conservative
group on Swansea Council; Conservative PPC for Swansea West
Conservatives, interviewed 12 March 2011.
Respondent Eight: Conservative PPC and MP for Worcester,
interviewed 28 March 2011.
Respondent Nine: Conservative European Election 2009 candidate
for the South East Region; Conservative PPC and MP for Suffolk
Coastal, interviewed 1 April 2011.

Publications

Carswell, D., and D. Hannan (2008) *The Plan: Twelve Months to Renew Britain* (Milton Keynes: Lightning Source).
CCHQ Wales (2009) *Welsh General Election 2010 Campaign Pack* (Cardiff: Welsh Conservatives).
Conservatives (2012b) *Campaign 2013 Toolkit* (London: The Conservative Party).
Conservatives (2012c) *Online Campaign Guide 2013* (London: The Conservative Party).
Dods (2010) *Dods Guide to the General Election 2010* (London: Dods).

Reports

IACC (2012) 'ICT on the Island', Report to the Meeting of the Board of Commissioners on the subject of ICT Services, Isle of Anglesey County Council, 13 February 2012, online: http://www.anglesey.gov.uk/x_links/14925, date accessed 23 July 2012.
Ofcom (2010) 'UK internet users becoming more security conscious', Ofcom, online: www.ofcom.org.uk/consumer/2010/05/uk-internet-users-becoming-more-security-conscious/, date accessed 23 June 2014.
ONS (2006) *Internet Access: Households and Individuals* (London: Office of National Statistics).
RWCA (2010) 'Annual General Meeting', Runnymede and Weybridge Conservative Association, Notes to accounts for the year ended, 31 December 2010.

Webpages

@DavidJonesMP (2014) David Jones, Twitter, online: https://twitter.com/DavidJonesMP, date accessed 21 June 2014.

DOI: 10.1057/9781137436511.0012

@JustineGreening (2014) Justine Greening, Twitter, online: https://twitter.com/JustineGreening, date accessed 21 June 2014.

@Mike_Fabricant (2014) Michael Fabricant, Twitter, online: https://twitter.com/Mike_Fabricant, date accessed 17 June 2014.

@RidgeNewman (2010) Anthony Ridge-Newman, Twitter, online: http://www.twitter.com/RidgeNewman, date accessed 22 May 2010.

Conservatives (2012a) 'MyConservatives.com', The Conservative Party, online: http://www.conservatives.com/ (home page), date accessed 23 September 2012.

Conservatives (2014) 'Sorry we couldn't find the page you were looking for: But now you're here, why not answer a few quick questions to find out how our long-term economic plan is helping you and your family?', The Conservative Party, online: http://www.conservatives.com/ (home page), date accessed 26 April 2014.

Owen, A. (2010) Albert Owen, webpage, online: http://www.welshlabour.org.uk/ynysmon/ (later http://www.albertownemp.org/), date accessed 14 April 2010.

Rees, D. (2010) Dylan Rees, website, online: http://dylanrees.plaidcymru.org/ (home page), date accessed 28 April 2010.

Ridge-Newman, A. (2010a) Anthony Ridge-Newman, website, online: http://www.RidgeNewman.com/ (home page) (later http://www.ridge-newman.com/), date accessed 22 May 2010.

Ridge-Newman, A. (2010b) Anthony Ridge-Newman, Facebook (politician page), online: https://www.facebook.com/anthonyridgenewman, date accessed 22 May 2010.

Websites

Blue Blog online: http://blog.conservatives.com/.

BNP online: http://www.bnp.org.uk/.

Conservative Party online: http://conservatives.com/.

ConservativeHome online: http://www.conservativehome.com/.

Facebook online: http://www.facebook.com/.

Gmail online: http://www.gmail.com/.

Guido Fawkes online: http://order-order.com/.

Iain Dale's Diary online: http://iaindale.blogspot.co.uk/.

Liberal Democrat Act online: http://www.libdemact.org.uk/.

JustGiving online: http://www.justgiving.com/.

DOI: 10.1057/9781137436511.0012

Mail.com online: http://mail.com/.
Mumsnet online: http://www.mumsnet.com/.
Sumall online: http://sumall.com/.
Tory Bear online: http://www.torybear.com/.
Total Politics online: http://www.totalpolitics.com/.
Twitter online: http://twitter.com/.
WebCameron online: http://www.webcameron.org.uk/.
Webs.com online: http://www.webs.com/.
YouTube online: http://YouTube.com/.

Secondary sources

Books and journals

Adoni, H., and S. Mane (1984) 'Media and the Social Construction of Reality toward an Integration of Theory and Research', *Communication Research*, 11(3): 323–40.
Anderson, L. (2006) 'Analytic Autoethnography', *Journal of Contemporary Ethnography*, 35(4): 373–95.
Anstead, N., and A. Chadwick (2009) 'Parties, Election Campaigning and the Internet: Toward a Comparative Institutional Approach', in A. Chadwick and P. Howard (eds) *Routledge Handbook of Internet Politics* (Abingdon: Routledge).
Arthurs, J. (2010) 'Contemporary British Television', in M. Higgins, C. Smith and J. Storey (eds) *The Cambridge Companion to Modern British Culture* (Cambridge: Cambridge University Press).
Bailey, R. (2011) 'What Took So Long? The Late Arrival of TV Debates in the UK General Election of 2010', in D. Wring, R. Mortimore and S. Atkinson (eds) *Political Communication in Britain: The Leader Debates, the Campaign and Media in the 2010 General Election* (New York: Palgrave Macmillan).
Bale, T. (2006) 'PR man? Cameron's Conservatives and the symbolic politics of electoral reform', *The Political Quarterly*, 77(1): 28–34.
Bale, T. (2008) ' "A Bit Less Bunny-Hugging and a Bit More Bunny-Boiling"? Qualifying Conservative Party Change under David Cameron', *British Politics*, 3: 270–29.
Bale, T. (2010) *The Conservative Party from Thatcher to Cameron* (Cambridge: Polity).

DOI: 10.1057/9781137436511.0012

Bale, T. (2012) *The Conservatives since 1945, The Drivers of Party Change* (Oxford: Oxford University Press).

Ball, S. (1994a) 'Local Conservatism and the Evolution of the Party Organization', in A. Seldon and S. Ball (eds) *The Conservative Century: The Conservative Party Since 1900* (New York: Oxford University Press).

Ball, S. (1994b) 'The National and Regional Party Structure', in A. Seldon and S. Ball (eds) *The Conservative Century: The Conservative Party Since 1900* (New York: Oxford University Press).

Ball, S. (1998) *The Conservative Party since 1945* (Manchester: Manchester University Press).

Ball, S. (2005) 'Factors in opposition performance: the Conservative experience since 1867', in S. Ball and A. Seldon (eds), *Recovering Power: the Conservatives in Opposition* (Basingstoke: Palgrave Macmillan).

Ball, S. and A. Seldon (eds) (2005) *Recovering Power: The Conservatives in Opposition since 1867* (Basingstoke: Palgrave Macmillan).

Bayard de Volo, L. and E. Schatz (2004) 'From the Inside Out: Ethnographic Methods in Political Research', *Political Science and Politics*, 37(2): 267–71.

Beck, U. (2002) *Individualization: Institutionalized individualism and its social and political consequences* (Sage: London).

Bimber, B. (1998) 'The Internet and Political Transformation: Populism, Community and Accelerated Pluralism', *Polity*, 31(1): 133–60.

Bourdieu, P. (1991) *Language and Symbolic Power, introduction and edited by J. B. Thompson* (Cambridge: Polity Press).

Burstein, D. (2005) 'From Cave Painting to Wonkette: A Short History of Blogging', in D. Kline and D. Burstein (eds) *Blog! How the newest media revolution is changing politics, business and culture* (New York: CDS Books).

Butler, D. and D. Kavanah (1992) *The British General Election of 1987* (London: Macmillan).

Chadwick, A. (2010) 'Britain's First Live Televised Party Leaders' Debate: From the News Cycle to the Political Information Cycle', *Parliamentary Affairs*, 64(1): 1–21.

Chang, H. (2008) *Autoethnography as Method* (Walnut Creek: Left Coast Press).

Charmley, J. (1996) *History of Conservative Politics 1900–1996* (London: Macmillan).

DOI: 10.1057/9781137436511.0012

Charmley, J. (2008) *A History of Conservative Politics since 1830* (Basingstoke: Palgrave Macmillan).

Clarke, C., and W. Kaiser (eds) (2003), *Culture Wars: Secular-Catholic Conflict in Nineteenth Century Europe* (Cambridge: Cambridge University Press).

Cockett, R. (1994) 'The Party, Publicity, and the Media', in A. Seldon and S. Ball (eds) *Conservative Century: The Conservative Party since 1900* (New York: Oxford University Press).

Cohen, A. P. (1982) *Belonging: Identity and Social Organization in British Rural Cultures* (Manchester: Manchester University Press).

Coleman, S. (ed.) (2011) *Leaders in the Living Room: Prime Ministerial Debates of 2010: Evidence, Evaluation and Some Recommendations* (Oxford: Reuters Institute).

Coleman, S., F. Steilbel and J. Blumler (2011) 'Media Coverage of the Prime Ministerial Debates', in D. Wring, R. Mortimore and S. Atkinson (eds) *Political Communication in Britain: The Leader Debates, the Campaign and Media in the 2010 General Election* (New York: Palgrave Macmillan).

Conroy, M., J. T. Feezell and M. Guerrero (2012) 'Facebook and political engagement: A study of online political group membership and offline political engagement', *Computers in Human Behavior*, 28(5): 1535–46.

Curran, J. (2002) *Media and Power* (London: Routledge).

Dahlgren, P. and Gurevitch, M. (2005) 'Political Communication in a Changing World', in J. Curran and M. Gurevitch (eds) *Mass Media and Society*, 4th edn (London: Hodder Arnold).

Davis, A. (2010) 'New media and fat democracy: the paradox of online participation', *New Media and Society*, 12(5): 745–61.

Dirks, K., and D. Ferrin (2001) 'The role of trust in organizational settings', *Organization Science*, 12(4): 450–67.

Dorey, P., M. Garnett and A. Denham (2011) *From Crisis to Coalition: The Conservative Party, 1997–2010* (Basingstoke: Palgrave Macmillan).

Downey, J. and S. Davidson (2007) 'The Internet and the UK General Election', in D. Wring, J. Green, R. Mortimer and S. Atkinson (eds) *Political Communications: The Election Campaign of 2005* (Basingstoke: Palgrave Macmillan).

Edwards, J. (2000) *Born and Bred: Idioms of Kinship and New Reproductive Technologies in England* (Oxford: Oxford University Press).

DOI: 10.1057/9781137436511.0012

Ellis, C., and A. Bochner (2000) 'Autoethnography, Personal Narrative, Reflexivity: Researcher as Subject', in N. Denzin and Y. Lincoln (eds), *The Handbook of Qualitative Research* (London: Sage).

Esser, F., C. Reinmann and D. Fan (2000) 'Spin Doctoring in British and German Election Campaigns: How the Press is Being Confronted with a New Quality of Political PR', *European Journal of Communication*, 15(2): 209–39.

Fisher, E. (2010) *Media and New Media Capitalism in the Digital Age: The Spirit of Networks* (Basingstoke: Palgrave Macmillan).

Fisher, J., D. Cutts and E. Fieldhouse (2011) 'Constituency Campaigning in 2010', in D. Wring, R. Mortimore and S. Atkinson (eds) *Political Communication in Britain: The Leader Debates, the Campaign and Media in the 2010 General Election* (New York: Palgrave Macmillan).

Fisher, J., E. Fieldhouse and D. Cutts (2013) 'Members Are Not the Only Fruit: Volunteer Activity in British Political Parties at the 2010 General Election', *The British Journal of Politics and International Relations*, online: DOI: 10.1111/1467-856X.12011.

Geertz, C. (1973) *The Interpretation of Cultures* (New York: Basic Books).

Gibson, R. (2013) 'Party Change, Social Media and the Rise of "Citizen-initiated" Campaigning', *Party Politics,* online: DOI: 10.1177/1354068812472575.

Gibson, R. K., A. Williamson and S. Ward (2010a) 'Whatever happened to the internet', in R. Gibson, A. Williamson and S. Ward (eds) *The internet and the 2010 election: Putting the small 'p' back in politics?* (London: Hansard Society).

Gibson, R. K., and I. McAllister (2013) 'Online social ties and political engagement', *Journal of Information Technology and Politics*, 10(1): 21–34.

Gibson, R. K., and S. J. Ward (1998) 'UK Political Parties and the Internet: "Politics as Usual" in the New Media?', *The Harvard International Journal of Press/Politics*, 3: 14–38.

Gibson, R., and S. Ward (2012) 'Political Organizations and Campaigning Online', in H. Semetko and M. Scammell (eds) *The Sage Handbook of Political Communication* (London: Sage).

Gibson, R., K. Gillan, F. Greffet, B. Lee and S. Ward (2013) 'Party organizational change and ICTs: The growth of a virtual grassroots?', *New Media and Society*, 15(1): 31–51.

Gibson, R., and M. Cantijoch (2013) 'Conceptualizing and measuring participation in the age of the internet: Is online political engagement really different to offline?', *The Journal of Politics*, 75(03): 701–16.

DOI: 10.1057/9781137436511.0012

Gibson, R., M. Cantijoch and S. Ward (2010b) 'Citizen participation in the e-campaign', in R. Gibson, A. Williamson and S. Ward (eds) *The internet and the 2010 election: Putting the small 'p' back in politics?* (London: Hansard Society).

Green, N. (2002) 'On the Move: Technology, Mobility, and the Mediation of Social Time and Space', *The Information Society*, 18(4): 281–292.

Gustafsson, N. (2012) 'The subtle nature of Facebook politics: Swedish social network site users and political participation', *New Media and Society*, 14(7): 1111–27.

Hammersley, M., and P. Atkinson (2009) Ethnography: Principles in Practice (Abingdon: Routledge).

Heppell, T. (2008) *Choosing the Tory Leader: Conservative Party Leadership Elections from Heath to Cameron* (London: Tauris Academic Studies).

Heppell, T. (2013) 'Cameron and Liberal Conservatism: Attitudes within the Parliamentary Conservative Party and Conservative Ministers', *The British Journal of Politics and International Relations*, 15(3): 340–61.

Heppell, T., and D. Seawright (eds) (2012) *Cameron and the Conservatives: The Transition to Coalition Government* (Basingstoke: Palgrave Macmillan).

Heppell, T., and M. Hill (2005) 'Ideological Typologies of Contemporary British Conservatism', *Political Studies Review*, 3: 335–55.

Hesse-Biber, S. and A. Griffin (2013) 'Internet-Mediated Technologies and Mixed Methods Research: Problems and Prospects', *Journal of Mixed Methods Research*, 7(1): 43–61.

Hill, M. (2013) 'Arrogant Posh Boys? The Social Composition of the Parliamentary Conservative Party and the Effect of Cameron's "A" List', *The Political Quarterly*, online: DOI: 10.1111/j.1467-923X.2013.02430.x.

Hill, A., L. Weibull and Å. Nilsson (2007) 'Public and Popular: British and Swedish Audience Trends in Factual and Reality Television', *Cultural Trends*, 16(1): 17–41.

Hindman, M. (2009) *Myth of Digital Democracy* (New York: Princeton University Press).

Hofstede, G., G. J. Hofstede and M. Minkov (2010) *Cultures and Organization: Software of the Mind: Intercultural Cooperation and Its Importance for Survival* (London: McGraw-Hill).

Howard, P. N. (2006) *New Media Campaigns and the Managed Citizen* (New York: Cambridge University Press).

DOI: 10.1057/9781137436511.0012

Howard, P. N. (2010) *The Digital Origins of Dictatorship and Democracy* (Oxford: Oxford University Press).

Jensen, M. J., and N. Anstead (2014) 'Campaigns and Social Media Communications: A Look at Digital Campaigning in the 2010 UK General Election', in B. Grofman, A. H. Trechsel, M. Franklin (eds) *The Internet and Democracy in Global Perspective* (Vienna: Springer).

Jones, S. H. (2005) 'Autoethnography: Making the Personal Political', in N. Denzin and Y. Lincoln (eds) *The Handbook of Qualitative Research* (London: Sage).

Kandiah, M. (1995) 'Television enters British politics: the Conservative Party Central Office and political broadcasting 1945–55', *Historical Journal of Film, Radio and Television*, 15(2): 265–84.

Kavanagh, D. (1972) *Studies in Comparative Politics: Political Culture* (Basingstoke: Macmillan Press).

Kavanagh, D. (ed.) (2013) *The Politics of the Labour Party, Routledge Library Editions: Political Science*, 55 (London: Routledge).

Kavanagh, D. and P. Cowley (2010) *The British General Election of 2010* (Basingstoke: Palgrave Macmillan).

Koot, W. C. J. (1995) *The complexity of the everyday: An anthropological perspective on organizations* (Bussum: Coutinho).

Kuhn, R. (2007) 'Media management', in A. Seldon (ed.) *Blair's Britain 1997–2007* (Cambridge: Cambridge University Press).

Lamprinakou, C. (2008) 'The Party Evolution Model: An Integrated Approach to Party Organization and Political Communication', *Politics*, 28(2): 103–11.

Lawes, C. and A. Hawkins (2011) 'The Polls, The Media and Voters: The Leader Debates', in D. Wring, R. Mortimore and S. Atkinson (eds) *Political Communication in Britain: The Leader Debates, the Campaign and Media in the 2010 General Election* (New York: Palgrave Macmillan).

Lawrence, J. (2009) *Electing Our Masters: The Hustings in British Politics from Hogarth to Blair* (Oxford: Oxford University Press).

Lawson, K. (1994) 'Conclusion: Toward a Theory of How Political Parties Work', in K. Lawson (ed.) *How Political Parties Work: Perspectives from Within* (Westport: Praeger).

Lee, B. (2014) 'Window Dressing 2.0: Constituency-Level Web Campaigns in the 2010 UK General Election', *Politics*, 34(1), 45–57.

Lievrouw, L. (2011) *Alternative and activist new media* (Cambridge: Polity Press).

DOI: 10.1057/9781137436511.0012

Lilleker, D. G. (2013) 'Empowering the citizens?', in R. Scullion, R. Gerodimos, D. Jackson and D. Lilleker (eds) *The Media, Political Participation and Empowerment*, (Abingdon: Routledge).

Lilleker, D. G., and N. A. Jackson (2010) 'Towards a more participatory style of election campaigning: The impact of Web 2.0 on the UK 2010 general election', *Policy and Internet*, 2(3): 69–98.

Lin, C. A. (2004) 'Webcasting Adoption: Technology Fluidity, User Innovativeness, and Media Substitution', *Journal of Broadcasting and Electronic Media*, 48(3): 157–78.

Livingstone, S. (2005) 'Critical Debates in Internet Studies: Reflections on an Emerging Field', in J. Curran and M. Gurevitch (eds) *Mass Media and Society* (London: Hodder Education).

Lofgren, K., and C. Smith (2003) 'Political Parties and Democracy in the Information Age', in R. Gibson, P. Nixon, S. and Ward (eds) *Political Parties and the Internet: Net Gain?* (London: Routledge).

Lusoli, W., and S. Ward (2004) 'Digital Rank-and-file: Party Activists' Perceptions and Use of the Internet', *The British Journal of Politics and International Relations*, 6(4): 453–70.

Margetts, H. (2006) 'Cyber parties', in R. Katz and W. Crotty (eds) *Handbook of Party Politics* (London: Sage).

McNutt, K. (2014) 'Public engagement in the Web 2.0 era: Social collaborative technologies in a public sector context', *Canadian Public Administration*, 57(1): 49–70.

Mergel, I., and S. I. Bretschneider (2013) 'A Three-Stage Adoption Process for Social Media Use in Government', *Public Administration Review*, 73(3): 390–400.

Mok, D., B. Wellman and J. Carrasco (2010) 'Does distance matter in the age of the Internet?', *Urban Studies*, 47(13): 2747–83.

Norris, P. (2000) *A Virtuous Circle: Political Communications in Post-industrial Societies* (Cambridge: Cambridge University Press).

Panebianco, A. (1988) *Political Parties: Organization and Power* (Cambridge: Press Syndicate).

Paris, C. M., W. Lee and P. Seery (2010) 'The role of social media in promoting special events: Acceptance of Facebook "events"', in U. Gretzel, R. Law, and M. Fuchs (eds) *Information and Communication Technologies in Tourism 2010* (Vienna: Springer).

Pich, C., D. Dean and K. Punjaisri (2014) 'Political brand identity: An examination of the complexities of Conservative brand and internal

DOI: 10.1057/9781137436511.0012

market engagement during the 2010 UK General Election campaign,' *Journal of Marketing Communications*, forthcoming.

Pickerill, J. (2003) *Cyberprotest: environmental activism online* (Manchester: Manchester University Press).

Ramsden, J. (1995) *The Age of Churchill and Eden, 1940–1957: A History of the Conservative Party* (Harlow: Longman).

Ramsden, J. (1996) *Winds of Change: Macmillan to Heath, 1957–1975: A History of the Conservative Party* (New York: Longman).

Schein, E. (2010) *Organizational Culture and Leadership* (San Francisco: John Wiley).

Scolari, C. A. (2012) 'Media ecology: Exploring the metaphor to expand the theory,' *Communication Theory*, 22(2): 204–25.

Seawright, D. (2013) ' "Cameron 2010": An Exemplification of Personality-Based Campaigning,' *Journal of Political Marketing*, 12(2–3): 166–81.

Segall, A. (2001) 'Critical ethnography and the invocation of voice: From the field/in the field-single exposure, double standard?,' *International Journal of Qualitative Studies in Education*, 14(4): 579–92.

Seldon, A., and S. Ball (eds) (1994) *The Conservative Century: The Conservative Party Since 1900* (New York: Oxford University Press).

Seymour-Ure, C. (1996) *The British Press and Broadcasting since 1945* (Oxford: Blackwell).

Seymour-Ure, C. (2003) *Prime Ministers and the Media: Issues of Power and* Control (Oxford: Blackwell).

Shirky, C. (2008) *Here comes everybody: The power of organizing without organizations* (New York: Penguin).

Siddique, S. (2011) 'Being in-between: The relevance of ethnography and auto-ethnography for psychotherapy research,' *Counselling and Psychotherapy Research*, 11(4): 310–16.

Smith, A. (2011) *Devolution and the Scottish Conservatives: Banal activism, electioneering and the politics of irrelevance* (Manchester: Manchester University Press).

Snowdon, P (2010b) *The Extraordinary Fall and Rise of the Conservative Party* (London: Harper Collins).

Snowdon, P. (2010a) *Back from the Brink: The Inside Story of the Tory Resurrection* (London: Harper Collins).

Southern, R. and S. Ward (2011) 'Below the Radar? Online Campaigning at the Local Level in the 2010 Election,' in D. Wring, R. Mortimer and S. Atkinson (eds) *Political Communication in Britain: The Leader*

Debates, the Campaign and the Media in the 2010 General Election (Basingstoke: Palgrave Macmillan).

Straw, W. (2010) 'Yes we did? What Labour learned from Obama' in R. Gibson, A. Williamson and S. Ward (eds) *The internet and the 2010 election: Putting the small 'p' back in politics?* (London: Hansard Society).

Tanis, M., and T. Postmes (2007) 'Two faces of anonymity: Paradoxical effects of cues to identity in CMC', *Computers in Human Behaviour*, 23(2): 955–70.

Taylor, A. (2002) 'Speaking to Democracy: The Conservative Party and Mass Opinion from the 1920s to the 1950s', in S. Ball and I. Holliday (eds) *Mass Conservatism: The Conservatives and the Public since the 1880s* (London: Frank Case).

Taylor, A. (2008) 'Preface', in T. Heppell, *Choosing the Tory Leader: Conservative Party Leadership Elections from Heath to Cameron* (London: Tauris Academic Studies).

Thakur, A. P. (ed.) (2006) *Weber's Political Sociology* (New Delhi: Global Vision).

Theakston, K. (2011) 'Gordon Brown as prime minister: Political skills and leadership style', *British Politics*, 6(1): 78–100.

Tsatsou, P. (2009) 'Reconceptualising "Time" and "Space" in the Era of Electronic Media and Communications', *PLATFORM: Journal of Media and Communication*, 1: 11–32.

Vaccari, C. (2010) ' "Technology is a commodity": the internet in the 2008 United States Presidential election', *Journal of Information Technology and Politics*, 7(4): 318–39.

van Dijck, J. (2012) 'Facebook as a tool for producing sociality and connectivity', *Television and New Media*, 13(2): 160–76.

van Dijck, J. (2013) *The Culture of Connectivity: A Critical History of Social Media* (New York: Oxford University Press).

Vitak, J., P. Zube, A. Smock, C. T. Carr, N. Ellison and C. Lampe (2011) 'It's complicated: Facebook users' political participation in the 2008 election', *CyberPsychology, Behavior and Social Networking*, 14(3): 107–14.

Ward, S. (2005) 'The Internet, E-Democracy and the Election: Virtually Irrelevant?', in A. Geddes and J. Tonge (eds) *Britain Decides – The UK General Elections 2005* (London: Palgrave Macmillan).

Ward, S., and R. Gibson (2003) 'On-line and on message? Candidate websites in the 2001 General Election', *The British Journal of Politics and International Relations*, 5: 188–205.

DOI: 10.1057/9781137436511.0012

Ward, S., R. Gibson and P. Nixon (2005) 'Parties and the Internet: an overview', in R. Gibson, P. Nixon and S. Ward (eds) *Political Parties and the Internet: Net gain?* (New York: Routledge).

Whiteley, P. F. (2011) 'Is the party over? The decline of party activism and membership across the democratic world', *Party Politics*, 17(1): 21–44.

Whiteley, P. F. and P. Seyd (1998) 'The dynamics of party activism in Britain: A spiral of demobilization?', *British Journal of Political Science*, 28(1): 113–37.

Whiteley, P., P. Seyd, and J. Richardson (2002) *True Blues: The Politics of Conservative Party Membership* (Oxford: Oxford University Press).

Whitty, M., and A. Joinson (2009) *Truth, lies and trust on the Internet* (New York: Routledge).

Williamson, A. (2010) 'Inside the Digital Campaign', in R. Gibson, S. Ward and A. Williamson (eds) *The Internet and the Election 2010: Putting the small 'p' back into politics* (London: Hansard Society).

Williamson, A., L. Miller and F. Fallon (2010) *Behind the Digital Campaign* (London: Hansard Society).

Wring, D. (2007) 'General Election Campaign Communication in Perspective', in D. Wring, J. Green, R. Mortimore and S. Atkinson (eds) *Political Communications: The Election Campaign of 2005* (Basingstoke: Palgrave Macmillan).

Wring, D., and S. Ward (2010) 'The Media and the 2010 Campaign: the Television Election?', in A. Geddes and J. Tonge (eds) *Britain Votes* (Oxford: Hansard Society).

Youngs, G. (2009) 'Blogging and globalization: the blurring of the public/private spheres', *Aslib Proceedings*, 61(2):127–38.

Conference papers

Cantijoch, M., D. Cutts and R. Gibson (2011) 'Participating in the 2010 UK E-campaign: Who Engaged, How, and with what Effect?', paper prepared for presentation at the ECPR General Conference 2011: *The Internet, Electoral Politics and Citizen Participation in Global Perspective*, Reykjavik.

Elder, C. (2010) 'General election 2010 – Action replay', paper presented at the Personal Democracy Forum.

DOI: 10.1057/9781137436511.0012

Gibson, R. (2010) ' "Open Source Campaigning?": UK Party
Organizations and the Use of the New Media in the 2010 General
Election', paper prepared for presentation at the Annual meeting of
the American Political Science Association, Washington DC.
Gibson, R., and I. McAllister (2011) 'A Net Gain? Web 2.0 Campaigning
in the Australian 2010 Election', paper prepared for presentation
at the 2011 Annual Meeting of the American Political Science
Association, Seattle, WA.

DOI: 10.1057/9781137436511.0012

Index

DOI: 10.1057/9781137436511.0013

DOI: 10.1057/9781137436511.0013

DOI: 10.1057/9781137436511.0013

DOI: 10.1057/9781137436511.0013